THE JEEP

History of a World War II Legend

THE JEEP

History of a World War II Legend

David Dalet and Christophe Le Bitoux

Schiffer

Translated from the French by Omicron Language Solutions, LLC.

This book was originally published under the title,
La Jeep: L'histoire d'une légende,
by REGI'ARM, Paris, France.

Library of Congress Control Number: 2013946378

Type set in JJStencil/Helvetica Neue LT Pro

ISBN: 978-0-7643-4460-2
Printed in China

Published by Schiffer Publishing, Ltd.
4880 Lower Valley Road
Atglen, PA 19310
Phone: (610) 593-1777; Fax: (610) 593-2002
E-mail: Info@schifferbooks.com

For our complete selection of fine books on this and related subjects, please visit our website at **www.schifferbooks.com**. You may also write for a free catalog.

This book may be purchased from the publisher. Please try your bookstore first.

We are always looking for people to write books on new and related subjects. If you have an idea for a book, please contact us at **proposals@schifferbooks.com**

Schiffer Publishing's titles are available at special discounts for bulk purchases for sales promotions or premiums. Special editions, including personalized covers, corporate imprints, and excerpts can be created in large quantities for special needs. For more information, contact the publisher.

THE JEEP

History of a World War II Legend

Photo Patrick Sarrazin

INTRODUCTION

Numerous politicians were photographed behind the wheel of a Jeep such as President Roosevelt in 1942.

Below: this photo, taken during an exercise in the USA, shows a column of jeeps comprising Bantam, Willys and Ford.

«*It first it was just a mutter, perhaps the sound of a motor that hummed in our ears ... Suddenly, around a bend, rather like a ghost, came a strange square car, without a windshield or a top, with a rectangular radiator grille and sharks' teeth.*

Inside were four soldiers, three of whom pointed their machine guns in all directions. The white star painted on the hood left us in no doubt of their identity and our hearts began to swell with joy."

This fine description, taken from Eddy Florentin's book Stalingrad in Normandy, sums up in a few lines, the intensity of the meeting between the French people and the Jeep and the formidable emotional impact it had on the population liberated during that memorable summer of 1944.

The large open "eyes" surrounding its gaping grille gave it a friendly look

and damped down its military image. Also its rather junior status and size compared to the GMCs, Dodge's, Diamond's and Brockway's that made up the impressive American armada made it everybody's favorite.

Its laid-back look, its shape, its sound and even its name became firmly fixed in people's minds. In short, if the idea had been to make the Jeep a marketing product, the U.S. Army couldn't have done better. With its famous white star, the Jeep quickly became the distinctive emblem of the liberating forces. Just one glimpse of this new star changed the feelings of the newly liberated population from anxiety to euphoria.

A Ford Jeep taking to the air during tests in 1941.

The Jeep was often used as an impromptu platform. Here, General Montgomery addresses British soldiers, who, in a few hours, will be landing in France.

That's how, in one season, the Jeep became the adopted child of millions of the French, who all saw it as the symbol of the freedom they had dreamed of for four long years.

Today, two generations later, if that strong feeling has dulled somewhat, it nevertheless remains in everyone's subconscious. The proof lies in the fact that in France the Jeep is, without doubt, the most collected vehicle and always gets a few amiable signs of gratitude whenever people see one.

Luckily, its formidable robustness has enabled it to cross the years without much damage, allowing us to come across numerous models superbly maintained and restored.

So, if the Jeep has won the undying affection of our elders for over sixty years now, it can still count on its seductive name and its classic look to charm the younger generation. We can only hope that this venerable sexagenarian's magic endures. It certainly deserves it.

BEGINNING OF THE LEGEND

September 1, 1941. A historical photo taken outside the Butler factory: the first Bantam.

The history of what would become the first Jeep began at the start of the 1930s in the United States. Pushed by the infantry and the cavalry, the U.S. Army, aware of being under-motorized, started looking round for a support vehicle that could transport light materials for the troops. The idea was to eventually replace the horse-drawn transport that was still largely in use in the American Army, and also to replace the sidecars whose useful space and displacement capacities were ridiculous for machines used in recon operations. In 1932 the Army's eye was drawn to the small Austin from Great Britain, built by American Austin.

The Austin was a small sub-motorized, two place pick-up that quickly showed its limitations on uneven roads and rough trails. No other serious pro-

This small pick-up, built by the American company, Bantam, was lent to the American Army for numerous tests.

The very first model of the "Bantam Reconnaissance Car Pilot." It was tested over 5,000 kilometers of intensive trials at Holabird Camp in October 1941.

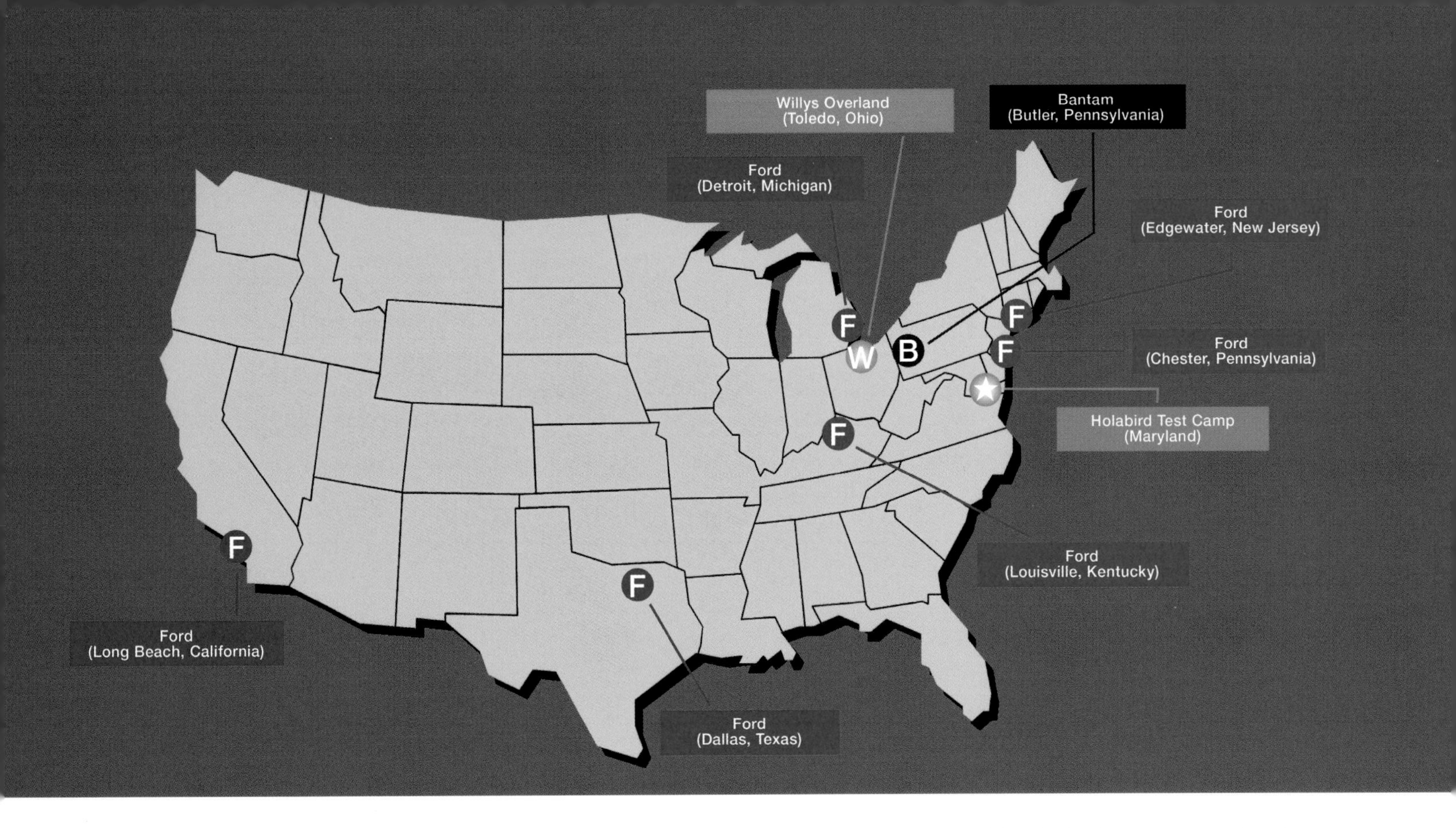

ject was considered: because of the isolationist policy, the lack of colonial interest and the emergence of their powerful economy, unnecessary expenditure was considered superfluous.

Things didn't really begin to pick up until the war in Europe in 1940 and the importance of light motorized vehicles became evident. So it was that in June 1940 a commission was put together within the U.S. Army to study the question. They were naturally interested in Bantam as this was the company who had the license to produce the famous little Austin. So the little Austin, now the Bantam, and the productivity capacities of the Butler factory were taken into account. The commission quickly realized that there wasn't an existing vehicle, including the Bantam, that would be suitable, even if it was modified and so in consequence a new vehicle had to be developed from A to Z, based on specifications put together by the

As the plate on the fender proves: the original Bantam prototype during tests at Holabird.

commission. The Secretary of State for War immediately gave the project the green light for a fleet of seventy vehicles. At the beginning of July Bantam proposed a contract to the U.S. Army for the construction and delivery of seventy pre-series vehicles. But as could be expected, the Quartermaster Corps refused to sign the contract as they were determined to bring competitors into the game and felt that the Butler factory was too small. The project was too important to leave in the hands of a single business of fifteen workers. Therefore, on June 27, 1940 the Quartermaster Corps launched a bid invitation for its vehicle to over 135 automobile companies throughout the country.

Of these companies, only two firms answered the invitation: Bantam, and one called Willys-Overland.

One of the seventy pre-series B60s that were ordered after successful trials at Holabird.

War Was Coming, They Had to Hurry!

The war in Europe was speeding up, so there was no time to waste! At Bantam and Willys-Overland everyone was on a war footing.

On July 22, 1940, Bantam, after an incredible race against the clock, sent the plans of its prototype to Holabird Test Camp in the suburbs of Baltimore, Maryland. Between July 17-22, Karl Probst, the chief engineer, worked night and day to create the plan of what would eventually become the Jeep! It was an incredible achievement, and although Willys-Overland offered a more attractive price, it quickly showed that it couldn't possibly meet the delivery deadline for elaborating the prototype and was certainly unable to deliver the pre-series vehicles in the seventy-five days demanded by the Army.

Therefore the U.S. Army bought the plans from Bantam and signed the contract on July 25.

Ford prototype. The bodywork was built by Budd, the coachbuilders. It is based on plans of the Bantam and the Willys.

The "QUAD" – Willys' first prototype. It is an exact copy of the Bantam BRC. The only real differences lie in the door openings and the wings.

On September 21, 1940, exactly two days before the fixed deadline for the prototype, the Bantam was finally ready.

Starting on September 27, the prototype was subjected to strict testing by the U.S. Army at Camp Holabird. The program comprised: crossing marshland, intensive cross country, bumpy roads, fording, driving through mud, dunes, hills and other steep ascents, endurance trials and sustained speed testing, all with added loads and towing different equipment of diverse lengths and weights.

At the end of these hellish trials, the general impression was good. The U.S. Army admitted that this was the best vehicle they had ever tested.

The Willys MA with its distinctive hood with the name stamped on it.

The Bantam BRC40, the last evolution made by their engineers. Its resemblance to the Jeep doesn't leave any room for doubt.

However, Bantam made notes of any faults that were noticed and the order for the seventy pre-series vehicles was awarded to them as the only company able to respect the deadline set in the bid invitation.

As to the other constructors and in particular Willys – the only other competitor – even though they hadn't produced a prototype, they were invited by the Quartermaster Corps to send observers to Holabird to watch the Bantam's tests. It was at that moment that another actor entered the stage: Ford, the automobile giant. Even though they hadn't even bothered to answer the initial invitation to bid, Ford intrigued by the interest in the new vehicle, sent their engineers to Holabird, the same as Willys.

Ford, drawn by the perspective of the prospective production, found at the same time, a way to feed its production line, which at the time, was underused. The Quartermaster Corps were obviously interested by the industrial potential of the American giant, which would not only decrease the delivery deadlines, but also would reduce the fabrication costs due to its assembly-line work strategy. Secretly, they allowed the Ford representatives consult Bantam's plans. Thus, the poor small Pennsylvanian factory, figuratively raped at its opening prom, lost the battle – the economic and strategic stakes were too high!

Side view of the BRC40. It is one of the last built and the last evolution before production ends, to the advantage of Willys and Ford.

During an exercise, this Ford Jeep is being tested for crossing gullies.

This Bantam BRC40 is one of the first order for 1,500 units which were delivered between March and June 1941.

In spite of the underhand way they had been treated, Bantam set to work to build the seventy pre-series vehicles, which were baptized B60 or Mark II, eight of which were four-wheel drive, and they were delivered on December 17, 1940. Even before the order was completed, the Quartermaster Corps, pressed by the war that was fast approaching the New Continent, but mostly because that believed that they had found in the Bantam, the basis of a fantastic vehicle, ordered 1,500 other

This Bantam four-wheel drive based on the BRC40 is part of an order for fifty units built at the end of 1941.

In 1941 Ford also built a four-wheel drive. But like the Bantam, production didn't go any further.

Bantam's for testing. Aware that the future quantities needed couldn't be carried out by a single constructor, and following their diversification politic, the Quartermaster Corps announced that they would be dividing the order between Bantam, Willys and Ford.

Finally, on November 11, 1940 – a month and a half after Bantam – Willys presented its prototype which had been perfected by their chief engineer Delmas J. Ross.

There's no need to point out that the vehicle was the exact copy of the original Bantam. Tests on the Willys-Quad (as it was called) began on November 13. The results were more than mixed: fragility and excessive weight. However, it was the powerful 60hp motor that saved the day, by giving the Willys-Quad stupefying performances on rough ground.

Ford, who had decided to join the adventure – pushed by the Quartermaster Corps and helped by having seen the Bantam's plans – also built a prototype named Pygmy. It too looked exactly like the Bantam; it was presented on November 23, 1940 at Holabird, two months to the day after the Bantam. Nevertheless, the vehicle was there and was tested along with the Willys-Quad, on the uneven roads at Holabird Test Center.

The results weren't entirely convincing, revealing the same weaknesses as the Willys and accentuated by a

From left to right: the Willys MA, the Bantam BRC40 and the Ford GP.

feeble tractor engine. However, the Quartermaster Corps, delighted that Ford with its huge infrastructure had joined the race, was careful not to knock anyone out of the competition.

Of these three test periods, it was objectively shown, after the weight problem was sorted out, that the best compromise was the Willys-Quad whose Go Devil motor saved the day and largely contributed to its success. At the end of the trials, each constructor placed a complete price offer.

Taking into account the two major measures, that is to say the quality and the cost, Willys was the objective choice. Not only had they presented the best vehicle, they followed this up by offering a highly attractive price.

Consequently, a first order of 16,000 examples of the series (MB) was given to Willys on July 31, 1941.

Head-on view of the Bantam BRC40 with its wrought iron radiator grille.

Three-quarter view of the Willys MA.

September 1943: three-quarter view, front and back of the Willys MB, final version.

On August 4, 1941, the order was upped to 18,600 examples.

During the time before starting on the order, the Willys engineers got together with the U.S. Army's engineers around the Willys MA (pre-series); the military engineers drew up an inventory of the parts and equipment needing to be standardized in line with the existing Army vehicles (air filter, shovel and axe, combat lighting) and other necessary modifications.

The new vehicle took the name MB (Military version B). Its production didn't start until November 18, 1941 due to diverse strike movements. It was planned to turn out 400 vehicles a day and the deadline was fixed for January 1942.

As for Ford, it began to re-establish itself, seeing that the Quartermaster Corps' initial argument about not putting all their eggs in one basket began to hit home with the War Department. Some considered the reinforcement of the

The Willys MA, easily recognizable because of its triangular reinforcements that give it better rigidity.

The Ford GP during tests in the summer of 1941.

Ford GP would complement the Willys MB and be able to answer a demand that could quickly become worldwide, seeing how the war was shaping up.

Finding themselves faced with this dangerous change of mind, Willys, afraid that it would follow in Bantam's footsteps, decided to preempt any governmental announcement. Trying to limit the damage, on October 14, 1941, Willys proposed an association with Ford and granted the government a non-exclusive license to allow Ford to manufacture the vehicles using Willys' specifications and plans. The Quartermaster Corps couldn't believe their luck and jumped on the chance to order 15,000 Ford MBs.

One can only try to imagine Bantam's indignation and fury when this was made public, it was the final humiliating blow, but their protestations were lost in the implacable mysteries of economics and geopolitics for two main reasons: the first came from preconceived ideas and prejudices; in fact, contrary to objective reality, few people believed in Bantam's capability to set up a real means of mass production.

The second reason, and certainly the most important, came from Japan who declared war on the United States on December 7, 1941, by attacking Pearl Harbor. From that moment onwards, everything that was important overrode anything else; including the furious protests made by Bantam, which in time became mere anecdotes.

The Bantam BRC40 with austere-looking seats.

Italy, September 20, 1943. General Mark Clark with Generals Teddar and House, drive through the town of Battaglia partially destroyed by Allied air forces.

ORIGIN OF THE NAME JEEP

November 1943: Major Griffith in his Jeep at the head of the convoy with a mechanics officer from the Chinese 22nd Division.

February 1944: Colonel Lloyd and two other soldiers stopped in front of the panel "Dobodura-Tokyo Road" in Oro Bay, New Guinea.

It's a vast program just trying, like many others before us, to give a precise and correct answer to this question. Because, it must be admitted, historians who have studied the problem are more or less stumped. The problem isn't finding an answer, but finding the right one. In fact, there's no shortage of ideas, even the most absurd.

The most serious referred to "Eugene the Jeep", Popeye's faithful companion. In fact, the term "Jeep" was first used in the media in an article by Katherine Hillyer from the Washington Daily News, which appeared on March 16, 1941, even though the real production hadn't yet started. It would therefore seem that the diminutive Jeep was originally given in 1940 to an artillery tractor produced by Minneapolis-Moline Company. In 1938, this firm based in Minneapolis, Minnesota, and which specialized in tractors and farm equipment, began to design a tractor intended for the Army.

Adopting the same politic as Bantam with the Pennsylvanian Civil Guard, Minneapolis-Moline lent the Minnesota National Guard four prototypes of their new tractor for life-sized trials. It was then that James T. O'Brian, a soldier with the 109th Ordnance Corps who had one of the tractors, nicknamed it "Jeep." He then made a comment about Eugene the Jeep, which had first appeared in the Popeye comics in 1936, drawn by Elzie Crisler Segar. The particularity of this personage was that he knew the answer to every problem and could do lots of incredible and surprising things. In the same way, O'Brian was amazed by the possibilities of this new vehicle that handled like a tractor, a truck and a tank. In short, like Eugene the Jeep, the tractor seemed to be able to resolve any problem and could do incredible things for the epoch.

Thus the word "Jeep" was entrenched in the Minneapolis-Moline Company and the Ordnance Corps when the vehicle was sent to the Test Center in Aberdeen, Maryland at the end of the summer of 1940. It rapidly became known as an amazing machine, accumulating superlatives and prowess. It was in another Ordnance Corps camp, at Holabird, that a vehicle was presented which would give all the meaning to the word Jeep: the prototype of the ¼ ton reconnaissance vehicle made by Bantam.

Word spread, as it had in the Ordnance Corps, and the press, impressed by the prowess of the small vehicle, baptized it Jeep in their articles.

November 20, 1944: the River Moselle in the worst flood for thirty years. The commanding general of the U.S. Army 3rd Corps circulates at the head of the column in the flooded streets of Pont-à-Mousson.

September 2, 1944, at Epinay Airfield. American officers in their Jeep observe a German Junkers Ju 88.

WAR PRODUCTION

Impressive, once loaded at the factories, the wagons usually head out for a port on the east coast to satisfy the help granted to the Allies in the vast lease-lend program.

In the factories, women workers often replaced the men.

Production of the MB started at Willys-Overland on November 18, 1941, while it started in February 1942 at Ford.

At Ford, it was the River Rouge factory in Detroit that launched the beginning of construction and assembly of the first GPW series on January 13, 1942. This activity ceased in River Rouge at the beginning of September 1942, so that they could concentrate on the construction of GPA's and trucks intended for the Army.

Willys and Ford: Who Did What?

That Willys and Ford were the only constructors of the Jeep is a well-known fact, but what exactly were the real tasks that fell to each of them during the marvelous Jeep saga?

In fact, at the beginning Ford had hired the Budd Company to build the GP bodies, as they had built the body of one of the two Pygmy's. However, this cooperation ended in October 1941 when the GP was no longer produced. After that, Ford built the bodies themselves

On Willys' side, limited by a small production capacity, they had to hire out the body construction throughout the production period. Ford, in spite of its huge industrial potential spreading over six sites, found itself in the same boat after January 1944, because its production line was monopolized by production for the war effort. In fact, looking closer, we see that the production of the GPW in the Ford factories represented a tiny part of the industrial activity within the American giant's factories. As well as the

Transport methods should be as rational, efficient and productive as the Jeep. Stacking filled all these requirements.

In one of the six assembly factories, the GPW comes off the production line in a frenetic rhythm, for the war effort. At this stage of the line, the vehicle moves under its own power. The next step is assembling the fittings and the final coat of paint.

277,896 GPWs and 12,778 GPAs and the civil production which continued, 8,685 B-24 Liberator bombers, 85,823 military trucks, 26,979 tank engines, 57,851 plane engines, 1,038 Destroyer M10 tanks and 1,690 Sherman tanks (to name but a few) were also produced in the Ford factories during the war period.

Taking advantage of the "composite body" that appeared in January 1944, Ford got in touch with the same supplier as Willys – the American Central Manufacturing Corporation (formerly Auburn Central Company), which was located in Connersville, Indiana. As for Willys – they had been using this tactic from the beginning of production of the MA pre-series!

Finally, in 1942, both Willys and Ford, finding themselves overwhelmed by the size of the orders and the set deadlines, were forced to order extra chassis from the A.O. Smith Company, based in Wisconsin. So, if the two constructors had to sub-contract the Jeep's two principle elements, they also had to do the same for the major part of the essential pieces of the rest of it.

To name only the principal players, Auto-Lite supplied most of the electrical equipment, the gear box came from Warner, the transmission and transfer case came from Spicer, the braking system was made by Bendix; Monroe supplied the shock-absorbers for Willys and Gabriel supplied them for Ford.

Thus, having more or less "planned" (thank you Bantam) the vehicle together, Willys Overland and Ford, faced with the enormous task, the necessity to lower costs and to ensure the best production rhythms, mostly turned to the same sub-contractors, who nevertheless respected the small differences and particularities of their clients' models. Actually, besides the symptomatic visible differences of the voluntary differentiation organized by Ford, the sub-contractors also had to make sure that the GPW's carried the famous "F" stamped on all their parts, even on the tires, unless Ford stamped them themselves when they received them.

Be that as it may, Willys and Ford were no longer, strictly speaking, the unique constructors but rather the giant assemblers. Setting up the production of this revolutionary small vehicle on such a huge scale widely overflowed the walls of our two manufacturers and involved a multitude of firms, both large and small, which, one by one, contributed their stone to accomplishing this formidable task. This was another thing that the Bantam Company found hard to swallow; in those circumstances, they could have swept away any doubts about their "limited" production capacity.

For extra space in the cargos, the Jeeps are partially dismantled before being packed in crates.

Theory and Reality: Two Separate Worlds

The complexities and requirements of mass production can sometimes lead to certain eccentricities, which once again, make the Jeep's history complex and subtle, far from widespread common stereotypes. There is no doubt that Ford, at the beginning of production in 1942, built their GPW's using the same chassis as Willys! Caught short by the size of the order, the strict deadline and with a production line that wasn't entirely operational, Ford, who usually built their own chassis (unlike Willys), couldn't supply the necessary quantities. Supplementary chassis were therefore ordered from the A.O. Smith Company in Wisconsin, who supplied Willys. The GPWs that came off the production

Lt. Walter J. Lind from San Francisco broke his right leg the day before the invasion. Treated on a hospital ship, he landed by Jeep on June 6, 1944.

PRODUCTION AT WILLYS OVERLAND:

Year	Production	Accumulation	Serial Number
1941	8 598	8 598	100 001 - 108 599
1942	91 423	100 021	108 600 - 200 022
1943	93 210	193 231	200 023 - 293 232
1944	109 101	302 323	293 233 - 402 323
1945	**57 542**	**359 874**	**402 324 - 459 874**

SIGNIFICANCE OF MB :

Following the **MA**, which were the initials for Military version **A** for the engineers at Toledo, MB simply means Military version **B**.

In the Coutances region, American officers watch a passing convoy of armored trucks advancing towards Brest. In the background is one of the many rail bridges destroyed during the invasion.

LA PRODUCTION CHEZ FORD MOTOR COMPANY

Year	Production	Accumulation	Serial Number
1941	-	-	
1942	87 573	87 573	501 - 87 574
1943	83 608	171 181	87575 - 171 182
1944	73 088	244 269	171 183 - 244 670
1945	**33 627**	**277 896**	**244 671 - 278 397**

SIGNIFICANCE OF GPW.

G: Identifies all government orders. **P:** internal classification at Ford identifying an eighty-inch wheelbase. **W:** the licence's origin, in this case Willys.

line therefore had all the standard GPW pieces, except for the chassis. Proof of this lies in a serial number stamped, following the example of all GPWs, on the front left-hand bar between the shock absorbers support and that of the motor. Besides, the serial number plate support belonging to Willys on the inside of the front left-hand side of the chassis had three pre-drilled holes but a plate was never riveted to it.

Production at Willys Overland

Not all the vehicles produced were intended for the American Army. In fact, of the 359,874 examples built by Willys, 348,849 were the result of contracts signed with the U.S. Government. So of the 11,025 examples left, it would seem that 11,000 were built for the Canadian Army and the rest were divided between a few government administrators and personal use and tests by personnel at Willys.

Production Site:

- Toledo, Ohio.

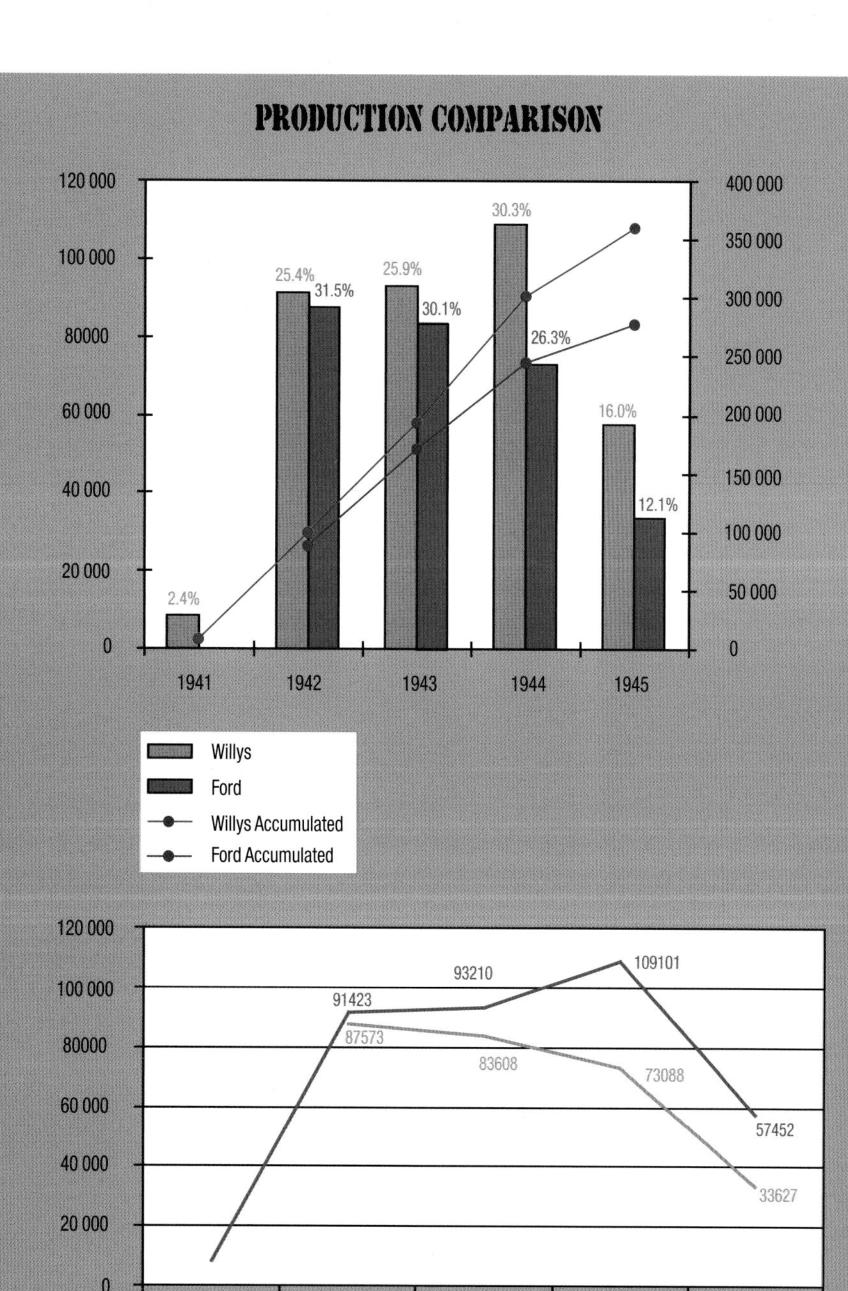

Somewhere in England, these soldiers are readying their Jeeps for the invasion. All the mechanical parts are carefully cleaned in petrol and checked. It's one of the numerous maintenance operations before the invasion.

Production at the Ford Motor Company

It is very hard, if not impossible, to get an honest production amount, as the information, albeit official, varies. In fact, whether it comes from Ford themselves or from their official historian during the Second World War, Mr. Lecroix, or from those attached to the specified quantities mentioned in the different contracts signed with the government, none of them are identical.

The numbers cited above are therefore a result of an evaluation, as close as it's possible to get, from the different sources consulted.

It should be noted that the production of GPWs was limited to the order placed by the American Army, for the simple reason that the free non-exclusive license ceded by Willys was restricted to the war period and orders placed by the American government only.

Thomas E. Mann of the 101st Airborne Division was a member of a three-man team, charged with repairing telephone lines. Here he is in the Haguenau sector after heavy snowfall. The Jeep is equipped with a type RL31 spool.

Production Sites:

- Dearborn, the River Rouge factory (Detroit, Michigan): Ford GP and GPW until 09/42.
- Chester, Pennsylvania
- Dallas, Texas
- Louisville, Kentucky

During the advancement, U.S. troops, in their Jeeps, enter a village in the Domfort region.

California, October 1942: This row of Jeeps ("peeps" according to the legend in those days) belongs to the 83rd Armored Scout Batallion of the 3rd Armored Division. Only the first Jeep is equipped with a radio. Note the nicknames stamped above the rear wheel.

- Richmond, California
- Edgewater, New Jersey

In view of these figures, one has to acknowledge that all the previously mentioned arguments to push Bantam out of the project are difficult to justify. In fact, concerning the first argument about Bantam's lack of productivity and the continued search for a heavy-weight in the matter – embodied by Ford, the figures show that during their respective production period, Willys, with their single production site, offered a better productivity with their 258 vehicles per day, than Ford, whose six sites only turned out 218 vehicles per day!

As to the second argument that questioned Butler's ability to increase its production capacity, the example of Willys speaks for itself. Willys, sixteenth national constructor in 1940, and deep in financial problems that put Bantam's in the shade, never-

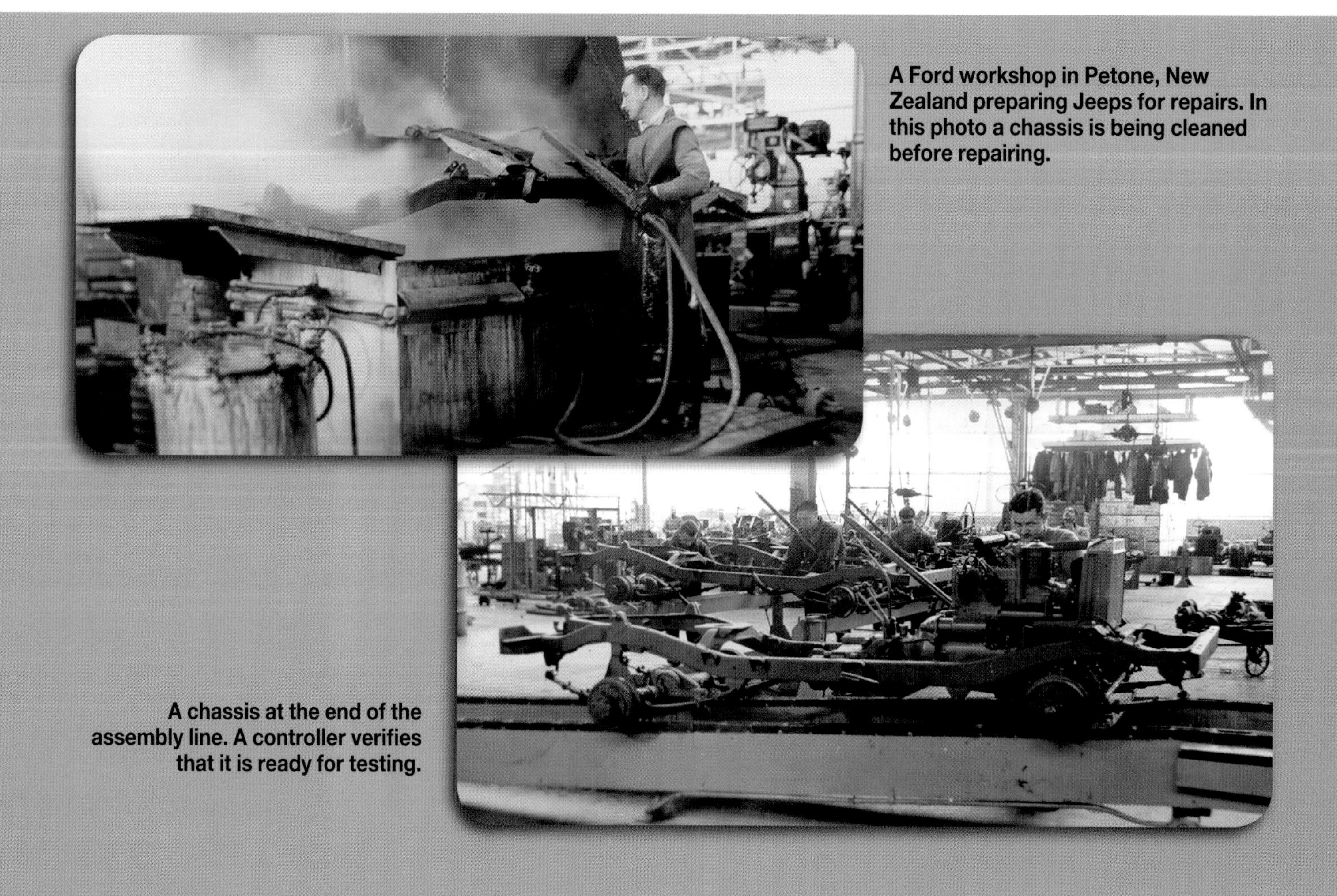

A Ford workshop in Petone, New Zealand preparing Jeeps for repairs. In this photo a chassis is being cleaned before repairing.

A chassis at the end of the assembly line. A controller verifies that it is ready for testing.

theless was easily able to quadruple its production capacity in its single Toledo factory in nearly two years, growing from an annual production of 26,000 vehicles to nearly 100,000 in 1943. Finally, to prove once and for all the injustice of the affair, it should be noted that a serious study carried out after the war, confirmed what the figures had already shown – Bantam had the capacity to produce the large quantities desired, thanks to its infrastructure, its knowledge and the size of its site, which had originally been built to turn out 184,000 vehicles a year in the time limit set by the American Austin Car Company. Actually, seeing the vast subcontracting that took place to produce the Jeep, these figures are undoubtedly pessimistic!

Belgium, January 12, 1945: A woman points out to American soldiers a panel on the wall of a house, indicating the retreat route taken by the German Forces.

A Jeep? You Want One? Here It Is!

Naturally, faced with the enormous need created by a world war, the Jeep wasn't limited to the American forces.

Thus, of the 637,770 MB and GPW Jeeps manufactured, it would appear that around 30% (182,597) were awarded to various allied countries in a vast lease-lend program. Among the main beneficiaries, the British Empire took first place with 104,430 Jeeps. The USSR received 49,250 vehicles, France 9,736, and finally China was given 6,944.

General view of an English port where the last LSTs that will be taking part in the invasion are being loaded.

TECHNICAL CHARACTERISTICS

JEEP TECHNICAL CHARACTERISTICS

Willys MB – Ford GPW

Motor :	Willys or Ford type Go Devil 6 volts, 116 4
Cylinders :	On line
Capacity :	2199cm3 (2.2l)
Compression ratio :	6.48/1
Capacity cooling liquid :	10.4L
Engine oil capacity :	4.73 l
Power :	60HP at 4000RPM
Max torque :	14.51 mf at 2000 RPM
Gear box :	Warner T84-J
Type :	shift stick, three gears and reverse.
Gears :	1st: 2665; 2nd: 1564; 3rd: 1; rev: 3554
Oil capacity :	0.71L
Transfer case :	Spicer 18
Type :	2 speed
Reduction ratio :	PV: 1.97/1, GV: 1/1
Oil capacity :	1.42L
Suspensions :	Suspensions:
Front :	Leaf springs
Rear :	Leaf springs
Shock absorbers :	4 hydraulic
Brakes :	4 hydraulic drum
Electrical equipment :	6 volts
Dynamo :	6 volts, 40 amps

Battery :	6 volts, 116 amps
Unloaded weight (without fuel or water):	1,060kg
Weight empty (ready to use) :	1,113kg
Length :	3.36m
Width:	1.57m
Height :	1.77m
Height windshield folded:	1.32m
Wheelbase :	2.03m
Tracks :	1.24m
Ground clearance :	22.23cm
Angle of attack :	45°
Output angle :	35°
Max speed :	105km/h
Tank capacity :	57L
Autonomy :	380km
Consumption on road :	13l/100km
Consumption off-road. :	18l/100
Slope max.. :	60%
Cant max. :	30%
Fording without preparation :	53cm
Tires :	600x16,6 ply
Number of seats :	4
Useful load :	363kg
Towed load :	454kg

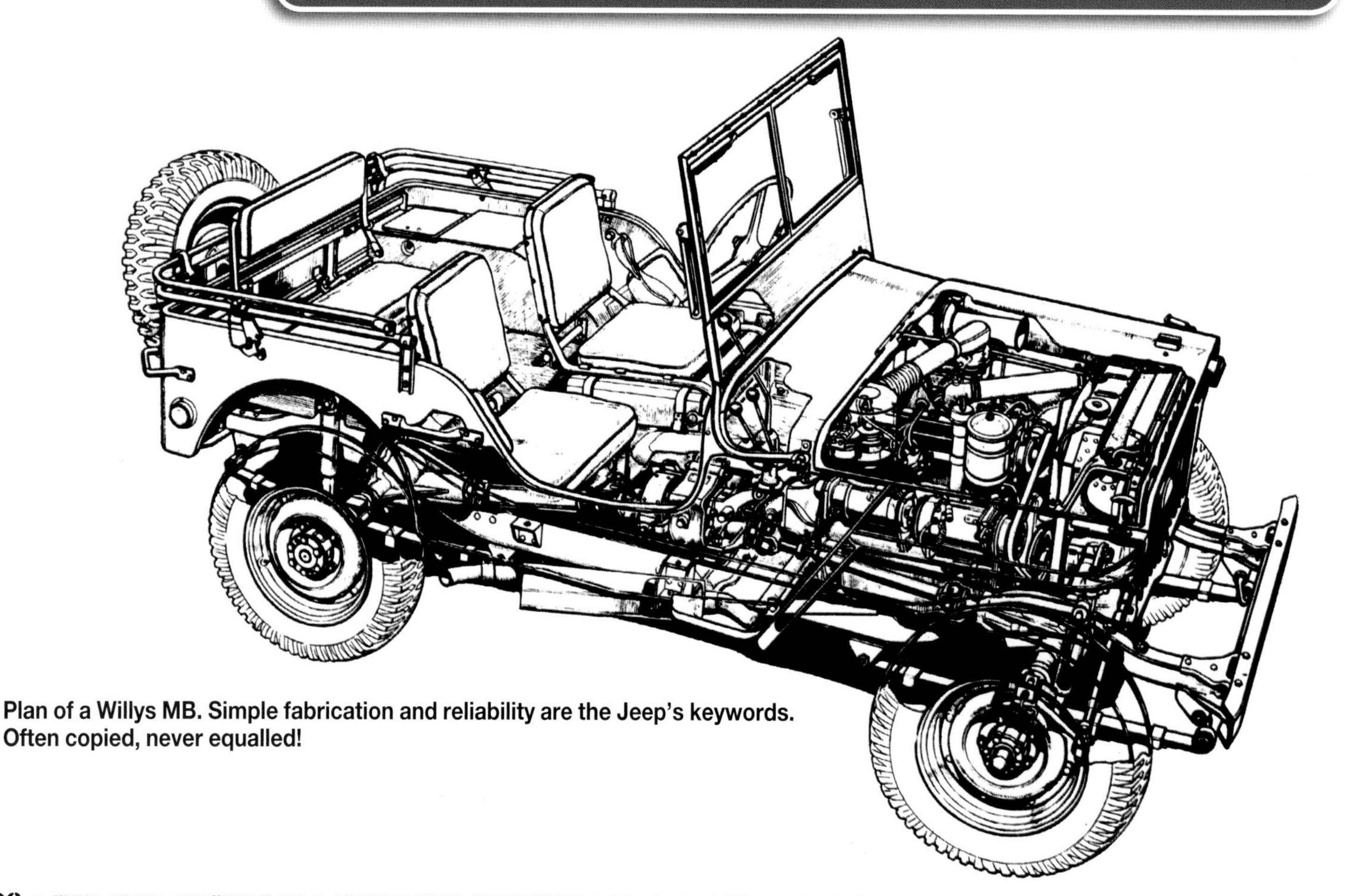

Plan of a Willys MB. Simple fabrication and reliability are the Jeep's keywords. Often copied, never equalled!

A unit of the U.S. 32nd Artillery Regiment disembark on the coast of Normandy.

STANDARD MODIFICATIONS BY DATE

"Back the attack" demonstration organized on September 27, 1943 in Washington. On this occasion, 400 American industrialists were invited to try out the Jeep.

1941	1942	1943	1944
	MECHANICAL		
December: modification of the back brake line moved towards the center	**May:** exhaust muffler replaced with a less noisy model. **June:** Front left suspension reinforced by adding a split leaf spring to compensate for the extra weight (driver and fuel reservoir) and the swing to the left when braking due to decentralizing the braking circuit.		**June:** reinforcement of the weight of the suspension's leaf springs.
	CHASSIS		
	February: modification of the battery's support.		
	WHEELS		
	February: full wheels changed for battle wheels easy to dismantle.		
	ELECTRICAL		
		January: key starter replaced by a simple switch. **March:** Addition of a radio connection box on the passenger's side. **December:** introduction of an ignition coil without earth connection on the bottom.	
	ENGINE		
	January: enlargement of the opening neck for refilling the engine oil. **March:** flat air filter replaced with a larger oil bath air filter.	**January:** modification of the engine oil gauge's stopper. **February:** modification of the crankcase's ventilation system from the inlet manifold.	

BODYWORK			
1941	1942	1943	1944
November: addition of a reinforcing bar and transversal air points under the front of the hood (abandoned in 1942).	**January:** the windshield's frame band increased by 5cm with the adaptation of fixing hooks on the hood. **February:** addition of a space-pocket with a key button. Brass windshield bolts were abandoned I favor of a steel bolt with a different shape. **March:** Willys replaced the wrought-iron radiator grille (slat grille) with a stamped slotted steel grille conceived by Ford. **April:** Lock on the glove compartment replaced by a push-button. **May:** The name "Willys" stamped on the left side of the rear panel on the body is abandoned. An oval, less noisy exhaust muffler replaced the round muffler. **July:** Two chain links are added to each side of the towing hook. **August:** Blackout lamps fixed to the front left wing. **September:** new reservoir receptacle with rounded edges. The filler neck is enlarged, which necessitates a modification of the driver's seat. **October:** Two triangular reinforcement plates are added to the sides of the tool case to rigidify the back panel. **December:** Installation of an air pump and its fixations under the back seat. During the year: - addition of a compensating leaf spring to the pack of springs on the front left side. - black Bakelite steering wheel changed for metallic three-branched one. - rubber windshield rests on the hood changed for wooden ones. - Leather shift gaiter changed for a rubber one.	**January:** introduction of the trailer socket, which necessitated repositioning the reflectors lower down. The spare wheel support with three pins is replaced by a two pin support with an added round plate. **March:** The old Ford and Willys stamped jerry can supports are replaced. **December:** The two triangular bodywork reinforcements with five holes and bearing the body's Serial Number are abandoned by Willys in favor for rounder models like Ford's, that have three holes, one of which is oval in the center. During the year: - rifle rack fixed at the base of the windshield. - first-aid box fitted under the dashboard.	**March:** Support for the lubrication pump fixed under the left side of the hood. For this the hood is reinforced with two rectangular sheet-metal plates. **During the year:** - appearance of vacuum operated windshield wipers. - decontaminator support placed under the passenger seat. - support for the lubrication plan placed under the hood at the end of 1944.

An interesting photo of a Jeep in February 1944. It has a wrought iron "slat" radiator grille, a blackout on the front wing and combat wheels.

Whether Ford or Willys, the name stamped on the rear panel disappeared in 1943.

The metal three-branch steering wheel appeared in 1942.

In August 1943 the blackout lamp was fixed to the front right wing.

In March 1942 the Jeep's radiator grille made of stamped sheet metal was adopted by Willys.

JEEP MARKINGS

The Jeep, like any other military vehicle, had symbols that had their own use, and just to make things difficult, often spoke a different language. Here is a detailed guide to the symbols most often found on a Jeep.

We start with the most visible and without doubt, the most emblematic: the star on the hood.

THE FAMOUS STAR ON THE HOOD CHANGED A LOT DURING THE WAR

From left to right: small and near the windshield during the campaign in Africa, it was enlarged later and placed in the center of the hood, circled in yellow so that Allied aviators could recognize it from above. From 1944 onward, and until the end of the war, the circle was changed to white and then was segmented. During the invasions of Italy and Normandy, M51 gas-detecting brown paint filled in the circle.

CODES	
0 :	Trailers and semi-trailers
00 :	Maintenance and breakdown vehicles
1 :	Liaison vehicles
10 :	Field kitchen
2 :	Vehicles up to 1 ton
20 :	Scout cars
3 :	Trucks up to 1 ton
30 :	Tanks
4 :	Trucks up to 4 tons
40 :	Tracked/semi-tracked trucks
5 :	Trucks over 5 tons
50 :	Fire trucks
6 :	Motorbikes and sidecars
60 :	Special vehicles (radio, etc.) and armored cars
7 :	Ambulances
70 :	Amphibious vehicles
8 :	Tractors
80 :	Tankers
9 :	Caterpillar tractors

A Jeep belonging to the 82nd Airborne Division drives through Pont l'Abbé village in Normandy.

It wasn't until after the campaign in Africa, and therefore in Italy, that the single small white star stamped on the hood was enlarged and set within a circle. This was done at the request of the Army Air Force so that they could more easily identify Allied vehicles. All the same, in Italy, the circle was yellow and was changed to white in England while preparations were underway for the invasion of Normandy. At first the circle was unbroken and then in 1944 it was segmented. In addition the unpainted part between the star and the circle could be painted brown with the famous anti-gas M51 paint (Liquid Vesicant Detector Paint M51). This paint had the particularity of turning red when it came into contact with liquid vesicant gases. It was mostly used on vehicles (Jeeps or others) on the frontline during large-scale operations, but it was only used for this purpose. When the windshield was folded down it was usually covered to avoid reflection; this cover had a star on it, either with or without a circle.

The registration numbers were painted in white on each side of the hood (black for the U.S. Marine Corps) and had seven, then eight, numbers.

These identification numbers were given to the Jeeps when they came off the production line and began with a two-number prefix, the 20, which identified the scout vehicles that the Jeep was a part of. The number following the 20 was literally the number in which it was recorded after being received by the military authorities (Government Serial Number), which had nothing to do with the chassis number showing the manufacturers' organization. So it was the only number and the only symbol that remained unchanged no matter what the vehicle was used for.

Following the custom established by bomber crews and fighter pilots, the Jeep quickly became a "graffiti wall" for the soldiers who drove them. They were often artistic and sometimes funny, but pin-up girls had an obvious advantage.

When it left the factory, these numbers were painted in bluish-gray and were quite small.

It was the only marking that the Jeep possessed at that stage. It was only when the vehicle was assigned to its future unit, and when it was received by the Army Administration Service on the assembly line (for Europe, these were mostly in Great Britain) that it was stamped with complementary symbols, one of which was the famous star on the hood.

Some, more anecdotic, were added in the regimental workshops of the vehicle's adopted unit. This dispersion of the markings explains the incredible diversity of the type and size to be found. They were either done using official or handmade stencils, or if time and manpower was short, they were hand painted. Be that as it may, the new number was strictly identical to the original, but was repainted in white. The new number was painted after the original one was removed, but sometimes it was painted over the original, which explains the superimposition of colors that can be seen on some vehicles.

From the beginning of the conflict and until 1942, the identification number was preceded by a "W" (War), indicating the U.S. War Department.

It was during the campaign in North Africa, which started in November 1942, that the Allied Forces (British, French and American) truly fought together for the first time with identical vehicles.

So that they could easily identify their vehicles, the American Army decided to replace the "W" with "USA" (United States Army), showing clearly the na-

TACTICAL ABBREVIATION	MEANING
Army Groups and Divisions	
GHQ	General Headquarter
AGP	Army Group
SHAEF	Supreme Headquarter Allied Expeditionary Forces
ASCZ	Advance Section Communication Zone
CZ	Combat Zone
CBS	Channel Base Section
NBS	Normandy Base Section
BBS	Britanny Base Section
LBS	Loire Base Section
DBS	Delta Base Section
OS	Oise Section
SS	Seine Section
CAS	Continental Advance Section
ABC	Antwerp-Brussels-Charleroi Highway Express
A	Army
AB	Airborne
AA	Anti-Aircraft
I	Infantry
AAF	Army Air Forces
△	Armoured Forces
H	Air force
Regiments and Battalions	
AAA	Anti-Aircraft Artillery
AMB	Ambulance
C	Cavalry
△	Armoured
E	Engineer
F	Field Artillery
G	Chemical
GI	Glider Infantry
I	Infantry
M	Medical
P	Military Police
Q	Quartermaster
ORD	Ordnance
R	Reconnaissance
S	Signal
T	Transport
TD	Tank Destroyer
X	Supreme Headquarter
BG	Bomber Group (Air Force)
FG	Fighter Group (Air Force)
Companies	
AMB	Ambulance
AT	Anti Tank
AM	Ammunition
DP	Depot
FLD	Field Hospital
G	Chemical
P	Military Police
MR	Mortar
M	Medical
RP	Repair
TN	Train
T	Transport
GAS	Gasoline transport

tionality and the vehicle belonging to it. These initials were only used by the Army and were changed to USN for the U.S. Navy and USAAF for the U.S. Army Air Force. These initials were stamped on the hood above the identification numbers, and sometimes the initials "US" were added to the passenger side of the windshield.

It isn't unusual to find an "S" added to the id number. This was reserved for vehicles built to receive a radio (usually 12-volts) and signified that the vehicle had been especially equipped to reduce interference.

Finally, concerning the hood, sometimes a yellow stamp bearing the words PRESTONE WINTERIZED or ANTIFREEZE followed by the year concerned (ex: PRESTONE 44). This indicated that the vehicle had been treated during the course of 1944 for freezing conditions (Prestone was the name of an antifreeze product). This inscription was usually stamped in yellow as were all the markings relating to maintenance and usage.

In this photo, taken at Fontainebleau, we can see the rather rare "shipping markings" at the base of the windshield.

Lastly concerning the stars, a small white star (a standard 15.2cm diameter), sometimes within a circle, was also stamped on each side of the vehicle between the reflector and the corner handle. Another star, usually 10cm diameter, was stamped onto the front fender on one side or the other of the crank hole. More rarely, a star was stamped onto the jerry can holder. Also rather rarely, a circled star the same size as the one on the hood was stamped onto the top for easier aerial identification. This was stamped only when the Allies had gained air supremacy (at the end of the war) as such a symbol was obviously double-edged.

The markings on the front and back fenders described the tactical situation of the vehicle within the Allied ranks. So the first element from left to right corresponded to the independent strategic unit (minimum: division) to which was added the unit the vehicle was assigned to, making up the second element. The third carried on this theme and usually indicated the company. The fourth element placed the vehicle exactly within the order of the company.

On the right side of a Jeep the shipping marking was usually on level with the front passenger's seat. This gave all the necessary details for transport by rail or by sea.

The registration number on the side of the hood is preceded by the famous "S" which indicated that the Jeep had received "anti radio parasite" treatment. The "TP35" indicated the pressure of the "military" tires. The markings on the jerry can indicated the liquid inside.

Also on the outer right hand side of the Jeep, on level with the passenger's footwell, or where the radio was eventually to be fitted, were the markings relating to the dimensions and weight of the vehicle: SHIPPING MARKINGS. They described the vehicle's bulk to help with loading procedures when they were moved by ship or by rail. They were also used on vehicles unloaded by TLC (Tank Landing Craft) or TLS (Tank Landing Ship) because of their "weak" tonnage. These markings were white as they had nothing to do with usage or maintenance.

Even though the shipping markings were stamped using different sized stencils, they always showed the following information: the length, the width, the height and the weight.

LTH (length) 11"1";
WTH (width) 5'2";
RED.HT (reduced height) 4'4";
WGT (weight) 2325LB (British)
1 TON 1 CWT (US).

In 1944 a fixed shipping plate was introduced, alongside the three instruction plates on the Jeep's dashboard between the glove compartment and the hand brake. This plate was intended to replace the shipping markings, with all the relevant information. The shipping markings were still used however, because they were easily seen without having to look inside the Jeep.

Usually on the four wings and/or on the dashboard, the tire pressure was indicated *TP 35* or *TP 35 LBS*, stamped in yellow as it was part of the maintenance instructions and was shown in pounds by square inch, the British pressure measure.

Sometimes, but rarely, precautions concerning the use of the transfer box were painted in yellow on the floor at the foot of the gear stick: *DON'T CIRCULATE ON HARD ROADS WITH THE FRONT AXLE GEARED.*

Concerning the windshield, along the bottom all sorts of names, nicknames, drawings, or short caustic phrases were added; during the war these became more and more widespread even though they weren't strictly allowed.

Sometimes the allusions were racy and daring and not at all in keeping with the conservative army way of thinking. This phenomenon is best illustrated in the French Army because of the accumulated rancor that inspired it. The most legendary was that of captain Dronne, commander of *la 9ème compagnie du RMT de la 2ème DB*, who was one of the first liberators of Paris on August 24, 1944. His first Jeep which became famous was christened, *"Mort aux cons" ("Kill the assholes")*. When it was destroyed in August 1944, its replacement was simply called *"Mort aux cons 2,"* and the third, just for a variation was called, *"Mort aux Boches."* Who would have thought of finding a vehicle with a name like that within the ranks of the cold and conservative French Army in the 1940s?

The markings on the hood were white. If they still existed they would show that the vehicle had passed through England before being dispatched to the Front, as the markings were done especially for the British population. They signified the presence of vehicles that weren't the same as the usual motors seen on British roads. The inscription on the back left hand side informed the public that the maximum speed was 40mph (this was sometimes reproduced in yellow on the dashboard for the driver).

The back right hand inscription read: CAUTION LEFT HAND DRIVE, with NO SIGNALS below it warning that the vehicle had no turning indicators.

Rather rarely, there might have been instructions about refilling the fuel tank, stamped in yellow on the reservoir: DO NOT OVERFILL ALLOW FOR EXPANSION.

The inscriptions on the jerry can served to specify the contents, the stock number, the production date and the

The Jeep's front fenders give all the information about its allocation number and its unit. From right to left we can see that it was the 28th Company D vehicle, of the 12th Infantry Regiment of the U.S. 4th Infantry Division.

The back fender has the same tactical information. For those Jeeps sent to England, standard security information was stamped on the canvas hood for the English population.

model that was stamped in yellow at first but later was changed to white. For the jeep, it stipulated that the gas used had an octane count of 80.

It should also be mentioned that according to U.S. Army standards, the lubrication points were painted matt red.

You often find fixed to the radiator grille, a disc indicating the vehicle's tonnage class. These discs were introduced in 1943 during the campaign in Italy and become obligatory in 1944. These tonnage discs usually comprised one or two numbers separated by a horizontal bar painted in matt black on a yellow background if the vehicle has a trailer. The upper number indicated the total tonnage of the vehicle and its eventual additions (trailer, artillery item).

The lower number showed the tonnage of the vehicle on its own. This information was intended for the Engineer's Units and the Military Police and was important for crossing bridges or when being transported by ship or by rail, for example.

In the case of the Jeep the tonnage class should be the same, whether the vehicle was alone or equipped with its famous standard trailer, either the Bantam T3 or the Willys MBT. The Jeep's tonnage class was 2 (2-ton), the disc shows two 2s separated by a horizontal bar. Where the Jeep had no attributed additions, the disc only had a single 2. In the British and the Canadian Army the Jeep's tonnage class was authorized to reach class 3.

Normandy 1944: this Jeep belonged to the 50th Armored Infantry Battalion of the 6th Armored Army, as we can see from the fender markings.

Therefore the tonnage disc showed a 2 (the Jeep's class) surmounted by a bar that in turn was surmounted by a 3 (the trailer's class). This disc was often placed in front of the front right hand headlight, hiding it completely (British and Canadian Army).

Having run through the paintings on a Jeep, it must be pointed out that it was very rare to find all these markings on a Jeep at the same time (the yellow markings to be clear!) So let's not overdo things, a Jeep was a war vehicle painted khaki and not a Christmas tree.

Navy Jeeps

Other than the characteristic black markings, the Jeeps allotted to the Navy stand out thanks to their charming blue-gray bodywork.

The registration numbers painted here and there on the hood were modified. The prefix 20 that indicated scout vehicles in the U.S. Army wasn't needed in the Navy. Therefore the registration number was reduced to five or six numbers surmounted or preceded by the corps it was assigned to, that is to say USN (United States Navy). Seeing that the Jeeps were mainly used for deliveries or transport, the tactical markings on the fenders were very simple: usually with the initials USN followed by the name of a ship or a specific service on the assigned base (USN TEXAS). Finally, a part from the black stars on the vehicle, there were no other noticeable changes, other than the star on the hood which had become random.

ON BOARD EQUIPMENT

March 1945 in Miesenheim, Germany. Two Jeeps pass each other, one equipped for transporting the wounded, the other with a luggage rack added on the back.

BLACK-OUT

The Jeep was equipped with two distinct blackout systems which permitted it to move in convoy but whose functions were different.

Understanding and Reading Blackout Messages

The first was achieved by the blackout headlight placed on the vehicle's front left wing. Covered by a hood to help avoid being detected by the enemy, its aim was to diffuse a thin ray of light. It had a double function. The first, in theory, helped the driver to see his way in the dark. The second, much more effective

For a distance less than 20m (60ft), two pairs of four lights indicate that the vehicle following is too close.

For a distance between 20m and 50m, two pairs of two lights indicate a correct distance between two vehicles.

From 50m to 250m one pair of lights indicates the presence of a vehicle ahead.

Invented for night convoys, the blackout lamp, with its thin beam of light surmounted by a visor, lights up the reflectors of the vehicle ahead. In this way the Allies' convoys could move at night without attracting the attention of enemy aircraft.

At the back, "cat's eyes" were integrated into the light fixture. In this particular case, it is the interior light transformed into an indicator that performs this function.

lit up the back reflectors on the vehicle in front enabling the vehicles to follow each other.

The second system, both complementary and different was represented by four blackout lamps also called "cats' eyes" (because they resembled the shape of the animal's pupils). Their use was to make travelling in column at night easier. This was done by changing the appearance of the lamps according to the distance between the vehicle and the observer.

For this, each of the two back blackouts were divided into four sections, grouped two by two.

The two front blackouts were used to locate the vehicle it was following and each of them was divided into two more or less equal sections.

At the front of the Jeep a pair of two lights indicates that the Jeep is less than 20m behind the one it is following.

For a distance superior to 20m, only single pair of lights can be seen.

Situated under the headlights, the "cat's eyes" signal the Jeep's presence to the vehicle preceding it.

SPADE AND AXE

Illogically the axe attached to the left side of the Jeep served no other purpose than to destroy the vehicle so that it could never fall into enemy hands. Obviously the soldiers used it for a lot of other tasks and rarely for its original purpose.

TOOL KIT

This was kept between the two standard toolboxes and contained the strict necessities for any minor repairs. Therefore it held six flat wrenches, multi-purpose pliers and an adjustable wrench. There was usually a screwdriver and a hammer too.

JERRY CAN AND SPOUT

The U.S. jerry can began to be manufactured at the end of the 1940s and was largely copied from the British model, which in turn had been copied from the German model, from which it takes its name (Jerry was a nickname given to German soldiers). In fact only the spout was different from the British and German versions. The U.S. model had a simple threaded cam instead of an integrated spout. It goes without saying that this wasn't very practical and a funnel was needed. The can also had a specific shape so that it could be easily stored without risk of losing it.

It isn't unusual to find a jerry can bearing a red plate with the octane content (80) stamped on it on level with the foremost side of the handle. When the octane count was 70, the plate was orange.

A jerry can for water appeared at the end of 1942; it had the same shape and dimensions as the gas jerry can, but there are two main differences.

The first is the stopper, which is a lot larger, like any opening that doesn't have threaded edges, because it has a lever system copied from the German model. The second difference lies in its conception; to be more hygienic, the inside was covered with oven-fired beige paint.

FIRST AID KIT

The first Aid kit for motorized vehicles comprises a box of twelve units. It first appeared in the Jeep in 1943 and was fixed under the dashboard with a special support between the handbrake and the glove box. A list of the contents was attached to the inside of the lid, along with instructions of the procedure to follow in case of burns, eye wounds, hemorrhage, open wounds, carbon monoxide poisoning, drowning or electrocution, fainting, broken bones or sprains.

FIRE EXTINGUISHER

This item was a powder projection model, operated manually by a pump situated on the top. It was housed on the driver's side under the dashboard, just behind the body's side reinforcement on a level with the clutch pedal. This space was taken into account by the constructors in the form of six holes necessary for fixing the support. In early models this space was in the same place but on the right side of the body.

M2 DECONTAMINATOR

Although luckily it wasn't needed during World War II, some Jeeps were equipped with M2 Decontaminator. This consisted no more nor less than a hand operated sprayer, somewhat like the fire extinguisher, which projected a fine lime chloride powder added to an emulsifying agent. Obviously this was intended to decontaminate the vehicle after a gas attack. At first the space and support for this item weren't in the plans. At the beginning of the campaign in Italy it was generally fixed to the inside of the windshield's metal frame. Then it was placed on the front wing mudguard. With the appearance of the "composite body" in 1944 the six holes needed to fix it under the passenger seat were added. The decontaminator was painted blue-gray or sometimes green.

This is the Jeep's traditional on board equipment, with the spade in its official cover and the axe below it. These tools were often used by the vehicle's crew for a variety of tasks.

The M31 support equipped with a .30 water-cooled machine gun.

A Jeep equipped with a radio during a parade in the U.S. in 1943.

WINDSHIELD AND HEADLIGHT COVERS

These items were used to cover the parts of the vehicle subject to reflecting the sun and alerting the enemy of its presence. It was put in place when the windshield was folded down.

Sometimes it had a star on it to replace the one it hid on the hood. The same equipment was used for the headlights and was a simple affair of two round hoods with laces to tie it to the headlight support.

CANVAS HALF-DOORS

This equipment, purely official, had in the field the same utility as the driving mirror cover – absolutely none at all! Its only use was to avoid being splattered with mud, but during a battle, a bit of mud was the last thing on a soldier's mind, and far from being sissies, the American soldiers preferred being able to get in and out of the Jeep with no problem.

If you don't know the official place of the first aid kit, you would waste precious minutes looking for it, even though its support is fixed with four clearly visible screws.

The jerry can could also be attached to the side of the Jeep.

Seeing the limited space inside a Jeep, the extinguisher is fixed next to the clutch pedal, which made changing gear difficult.

At first the Decontaminator M2 was attached to the wing's mudguard. Later it was placed under the front passenger seat. Luckily for the Allies it wasn't needed during the war.

The capstan was fixed in front of the radiator grille. Often used in place of a winch, it was intended for Navy Jeeps.

A Jeep being used as a tractor for the quadruple .50 M55.

CANVAS WATER BAG

Unlike the half-doors, this equipment given out from 1943 onwards was much appreciated and it showed. The presence of this foldable, waterproof canvas bucket, usually tucked behind the jerry can's straps, can be seen on numerous vehicles. It was used to stock water for multiple uses – from a morning wash to potato peeling duty.

AIR PUMP

This appeared at the end of 1942 and in spite of the interest in the T1 compressor, the air pump was never replaced. Its advantages included its price, its weight and its simple maintenance requirements. Being a part of the standard on-board equipment, like the starting handle, the lubrication pump or the tool kit, it was to be found in a large number of Jeeps. But unlike its counterparts, it had its own space especially provided for it. It was placed under the back seat and had an adapted support. It also had an adapter shaped like a buzzard enabling it to be used to clean out pieces such as the carburetor.

LUBRICATION PUMP

This was used to grease the necessary parts during the usual and regular maintenance of the Jeep. There were two models; the first, made by Alemite with a longitudinal pumping system, was housed in one of the tool cases in the back of the vehicle. A second model, more difficult to find these days, appeared at the beginning of 1944; it had a side lever pump system, more efficient and less difficult to use. This model, larger than its predecessor was kept in place by a support under the left hand side of the hood.

OIL CAN

This was fixed to its support in the engine compartment, behind the carburetor and to the left of the horn.

COMBAT WHEELS

Like the wrought-iron radiator grilles, the Jeep had an early model of wheel that quickly gave way to a second more practical type. So after the first 20,000 vehicles had been equipped with wheels cast in one piece, the Jeep adopted removable wheels in two sections. They could be removed with a simple No.19 wrench, in order to change the tire or the inner tube when it was punctured. Even the name "combat wheels" made a direct allusion to the fact that it could even be dismantled and repaired on the battlefield without tools or specialized personnel.

THE TIRES

The Jeep's tires produced during the war had a "military" structure and were mostly fabricated by Goodyear

Mostly used for washing, the canvas water bag was often attached to the jerry can

Half-door in place.

or Firestone. Just a small detail: Willys mostly subcontracted with Goodyear for their tires, while Ford who owned its own tire factory, fabricated their own which were stamped with their initials on the side. However, from 1944, after Ford Junior married the daughter of the director of Firestone, this latter won the contract to supply all the tires for the GPW.

Concerning the dimensions, they were standard on the Jeep, that is to say: 6.00x16 Ply: 6.

6.00: thickness of the tread and the height of the sides in inches; 16: the diameter of the interior of the tire and the exterior of the wheel rim in inches; Ply: 6: thickness of the internal framework (6 means 6 juxtaposed layers of fibers).

SPECIAL EQUIPMENT

TOW BAR

This piece of equipment appeared in January 1943 and was called *Tow Bar Field Kit for Tandem Operation*. As its name implies, this triangle allowed the use of Jeeps in tandem to tow heavy pieces of artillery, such as the well-known 105mm howitzer. In this way two Jeeps harnessed together were able to replace a GMC or any Dodge as an artillery tractor.

This kit could be installed in an hour and replaced the fender. A reinforced plate on the back hook was intended to compensate for the considerable effort exercised on the hook by the consequent weight of the artillery towed.

Finally, when not in use, the bar was strapped to a small wooden base fixed to the radiator grille.

THE CAPSTAN

A certain number of Jeeps were equipped with a capstan for winching. The model used was simply the one that the Jeep GPA was equipped with. It was installed on the front of the chassis and could tow a weight of over 2,000kg.

DEEP FORDING KIT

The Jeep couldn't cross water deeper than 50cm. Any deeper and risks of

From 1942 onwards, the air pump was placed under the back seat and didn't take up much space.

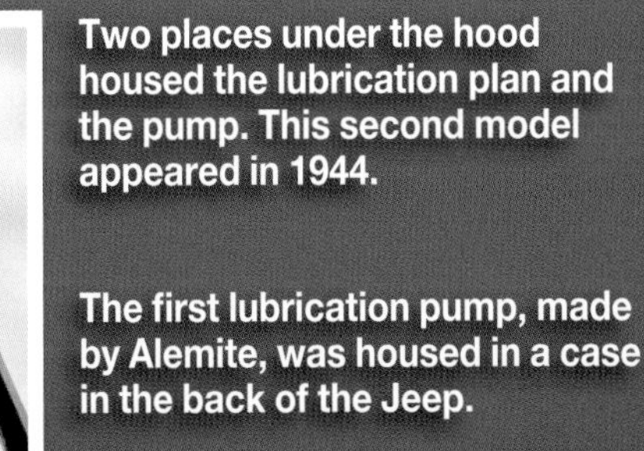

Two places under the hood housed the lubrication plan and the pump. This second model appeared in 1944.

The first lubrication pump, made by Alemite, was housed in a case in the back of the Jeep.

The "Tow Bar Field Kit for Tandem Operation," or simply tow-bar, gave an accumulation of traction power from two Jeeps, principally for artillery items, but also for other uses.

The famous combat wheel's main specificity was that it could be dismantled on the battlefield with a simple No.19 wrench. It is immediately recognizable because of its eight bolts.

breaking down were almost certain; the need to add ad hoc equipment was the only solution to be able to take part in landing operations, for example.

The driver and every electrical part were carefully waterproofed with a thick layer of Asbestos. All mechanical parts that wouldn't work in water received the same treatment. The protection had an air inlet and an outlet for exhaust fumes in case the motor stalled. The air inlet was achieved by a supple rubber tube attached to the carburetor under the hood, coming out on the right hand side of the windshield. For the exhaust system, the tailpipe was dismantled and replaced by a metal tube, directly fixed to the converter and which led to the left hand side of the windshield frame. This long and tedious conversion allowed the Jeep to drive through water up to 1.50m deep, but only for about eight minutes.

These transformations were regularly used during landings in Sicily and Normandy, in French Provence and the hard campaign in the Pacific.

CABLE CUTTER

This homemade addition was screwed or welded onto the front fender and could be reinforced with one or two arms. Its use spread rapidly after the campaign in Italy because the Germans had the unfortunate habit of stretching a rope, often a metal one, across the road usually just round a bend, knowing that the Americans drove with the windshield folded down to avoid reflection from the sun giving away their position. It's easy to understand that this simple addition was highly popular as it meant that the Jeep's occupants could literally keep their head on their shoulders!

LUGGAGE RACK

This was also homemade and gave extra space for transporting additional equipment when necessary in the Jeep. When you think of the space of this latter compared with a GI's average pack, you could say that such equipment wasn't often a luxury.

EXPANSION TANK

The first models were called *"desert cooking kits"*, that is to say a cooling system adapted for the high temperatures in the desert.

The cable cutter was a homemade item that was welded or bolted to the Jeep's fender. An extra arm or two could be added to reinforce it.

The Jeep was fitted out to receive various types of radio, such as the SCR610 kit. This comprised a BC659 radio set on its PE120 box that was fixed to an FT250 support. The TS13 receiver was placed next to the AN29C foldable antenna when being transported.

Certain Jeeps had a luggage rack – the best way to increase the rather small space inside the vehicle. It could take on different forms, according to needs in the field.

The Deep Fording Kit meant that the Jeep could drive through water up to 50cm deep. On the right side it was equipped with a flexible tube for ventilation and on the left with a metal tube to evacuate exhaust gas.

This photo shows Jeeps equipped with the Deep Fording Kit, disembarking from an LCT during an exercise on a particularly inhospitable beach.

The second models were reserved for Jeeps that had over-heating problems in temperate areas.

For high temperatures, especially in the Mediterranean area, the Jeep was fitted out with an expansion tank. There were two "desert cooling kit" models. Above: the "Sahara" model was very voluminous.

This is the "Mediterranean" model, much smaller.

If the system of these two kits was identical, their size wasn't the same: the *"Sahara"* model was three times larger than the *"Mediterranean"* model.

Whatever the model, there wasn't a better cooling system. Its system of recuperating steam meant that less water was wasted refilling the radiator. So instead of water evaporating out of the radiator's overflow vent, the steam, recuperated by the expansion tank fixed to the radiator grille, once it became water again was taken back into the radiator, meaning that the necessary circuits never lacked water.

RADIO

According to the unit it was assigned to, a Jeep was equipped to receive a radio. The most common were the send/receive BC659 models that had a 15km range. They were usually attached to the back left-hand wing by a FT250 support and fed by 6- or 12-volts from its PE120 box onto which it was fixed. In the first Jeeps, the radio was fed by a long cable that ran through the middle of the vehicle and was attached to the battery. In January 1943, the radio was linked to a terminal box placed inside the bodywork on the passenger's side.

WEAPONS

As a first class multi-function vehicle, from the beginning the Jeep was naturally equipped with all sorts of weapons.

It had a rifle rack behind the windshield. Besides, its chassis had a reinforced metallic support behind the front seats. Two pedestal models were especially developed: The *Truck Pedestal Mount M31* that could be installed on the reinforcement on the chassis. This single footed attachment was adapted for the .50 or .30 machine-gun or the caliber .30 rifle (BAR).

The *Cal .30 Machine Gun Mount M48* could be installed in the place of the conveyer. Other installations were also tried out without any of them becoming a standard type, like the rack fixed on the right hand side of the dashboard or the circular rack allowing 360° movement or the gun mount for the twin-tureted anti-aircraft .50 machine gun.

From the start, the Jeep was used to tow the 37mm anti-tank gun. Tests were carried out to see if the same gun could be implanted directly in the back of the vehicle.

A lot of improvisations saw the light of day to support all sorts of weapons, for example mounting two .50 bazookas instead of one on a M31 support.

A Jeep equipped with a 4.2 inch mortar.

The fire support version of the Jeep, equipped with a twelve-tube rocket launcher.

The 37mm M4 cannon is one of the rare artillery pieces that could be towed by a single Jeep. Its small caliber meant that it was soon replaced, but it was often seen behind a Jeep during exercises in the U.S.

Here, the *Truck Pedestal Mount M31* is equipped with a .50 machine gun.

Not forgetting the specific applications such as the British, French and Belgian SAS Jeeps that were produced in limited numbers.

They also tried to get the Jeep into the heavy weapons brigade. A lot of important tests were carried out to install a 4.2 inch mortar, but they weren't followed up.

A small number were equipped in the back of the vehicle with a support for a twelve-tube 4.5 inch rocket launcher with protection for the windshield and drivers' seat.

The Jeep was also tough enough to tow a quadruple M55 support with the magazines installed in the place of the back seat.

In this photo, the rifle rack, useful for safely stowing a weapon, houses a .30 caliber M1 rifle.

France, September 13, 1944. This Jeep is fitted out as a railroad vehicle so that the engineers could check out the state of the tracks. The road wheels are attached to the sides. These Jeeps were particularly used in Asia.

THE FORD GPA

Completely fitted out, the amphibious Jeep (or Seep) could carry five passengers. The users' instructions were all over the dashboard, in the hope that it didn't sink!

The idea of an amphibious vehicle that would be equally at home in water as on the road had long been the dream of both the army and automobile constructors. So from the 1940s onward a number of tests were carried out and a number of prototypes were produced in a lot of countries. Unfortunately, none of them passed the tests. It is easy to imagine the technical problems the automobile industry faced. The major problem was the hull's water-tightness, but also the different exits necessary for the transmission axle or the screw shaft. Another delicate point was the weight of the material that would weigh the vehicle down – according to Archimedes' theory. In short, the battle wasn't won, but the arrival of the Jeep only made the project more tempting and realizable given its technical prowess, so it was susceptible to represent the amphibious vehicle's future. Therefore, once again, it was the Quartermaster Corps that launched the project in April 1941 via their Motor Transport Engineering Branch (a department that studied motorized vehicles) at Holabird. The hull's concept was naturally given to Sparkman and Stephen's New York-based naval shipyard, while the mechanical adaptation was taken on by Marmon-Herrington in Indianapolis, who since 1931 had specialized in integral traction and mass-produced vehicles. In this line, Ford was its preferred partner. So it was obvious that the Indianapolis Company turned to them for the Jeep's mechanical parts that were produced in the Ford factories. The project advanced and seeing the dependence on the original Jeep parts, it was finally Ford who took over the project. Therefore the finalized vehicle was called the GPA, after Ford's nomenclature: G (Government) that corresponded to all fabrication intended for the country; P for the ¼ ton vehicle's program and A for amphibious.

A first contract for seven 896 GPAs was signed on April 10, 1942. A total

Three-quarter front view of the 3rd Amphibian-3 prototype driven by Ford factory workers. Like the first series of conventional Jeeps, the Ford logo is stamped on the bodywork, in this case, on each side of the hull.

DATA SHEET

Motor	GPW 2.2l (Go Devil) 60HP
Brakes	Hydraulic drum
Tires	6.00x16
Max speed	95km/h
Max speed in water	8.5km/h
Autonomy	380km
Autonomy in water	60km
Max slope	45%
Approach angle	37.5°
Exit angle	37°
Electrics	12 volts (2x6V)
Length	4.63m
Width	1.63m
Track	1.24m
Height	1.73m
Ht. windshield folded	1.36m

Land configuration of this early GPA, the 225th example built as its Registration Number 702328 shows. The motor's ventilation trap is open as advised in the technical manual.

The GPA is a real Jeep, or rather a funny little boat on wheels. Wanting it to do everything, the amphibious Jeep never had the polyvalence hoped for, but it was useful for fording a large number of rivers.

The famous Go-Devil shows that it can get its sea legs at any given moment. It houses the capstan's transmission mechanism, which is directly connected to the crankshaft pulley.

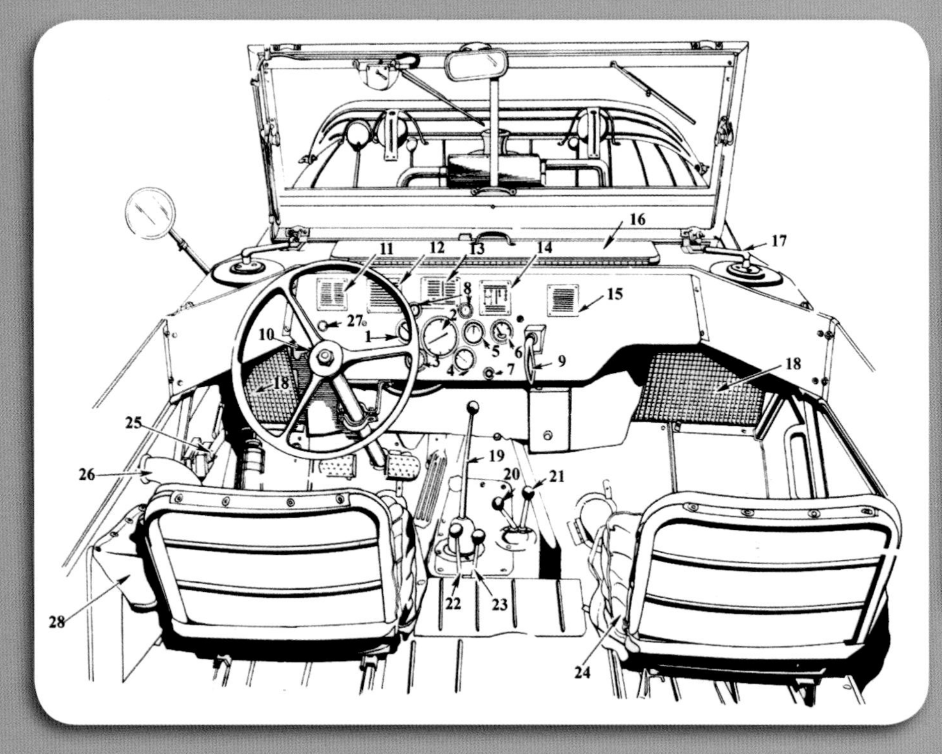

1. Gas gauge. 2. Speedometer. 3. Oil manometer. 4. Motor temperature gauge. 5. Ammeter. 6. Voltmeter. 7. Capstan clutch. 8. Dashboard light. 9. Handbrake. 10. Headlight switch. 11. Identification plate. 12. Transmission box plate. 13. Nautical instructions plate. 14. Plan of lever command plate. 15. Electrical instructions plate. 16. Ventilation trap when in water. 17. Opening/closing exterior ventilation command. 18. Ventilation grille when in water. 19. Transmission stick. 20. Front axle engagement lever. 21. Reducer engagement lever. 22. Propeller engagement lever. 23. Hold-pump engagement lever. 24. Pillow buoy. 25. Front deck trap command lever. 26. Hold-pump evacuation tube. 27. Holdall.

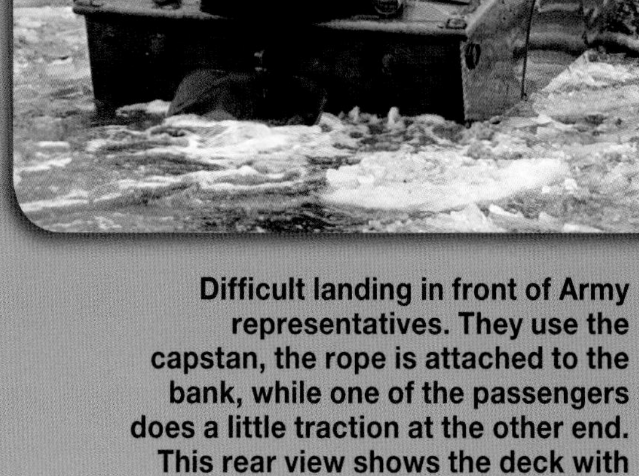

Difficult landing in front of Army representatives. They use the capstan, the rope is attached to the bank, while one of the passengers does a little traction at the other end. This rear view shows the deck with the protection grille under which the exhaust pipe exit.

Third period for this Ford GPA, the 12085th vehicle built as indicated by its Registration Number 7014189. It has reached its maximum evolution: gaff positioned on the right hand side, exhaust muffler protection, rifle rack and anchor fixed in their official places on the spare tire.

The GPA's onboard equipment included an anchor, a rope, two oars housed in the cockpit, and a pole fixed on the right hand side.

Seen from this angle the GPA shows off its propulsion tunnel that protects the propeller. The rudder is combined with the steering wheel.

Underside view of the Amphibian-2. The sheet metal fixed by hinges, sealing the propeller tunnel, is a simple protection while on the road. In the water it folds back automatically under the pressure from the propulsion.

The same vehicle being presented to a group of officers from diverse military branches as we can see by their uniforms.

March 13, 1944: somewhere in England in one of the multiple U.S. Ordnance depots. 2nd and 3rd generation GPAs are being regrouped for the approaching invasion.

of 12,778 GPAs came off the Ford assembly line between September 1942 and June 1943, a production time of ten months. The modest number and early end of production was caused by the fact that the initial enthusiasm quickly died out when faced with the terrain in reality. The GPA disappointed because, trying to do everything top class, it did everything in an average manner. Its buoyancy was questionable and useless for use in the sea, as it had a tendency to settle low in the water because of its weight. Its abilities on rough land weren't much better because it was seriously restricted by an additional encumbering weight of 606kg and its overall shape was little disposed to keeping all four wheels on the ground or to attack and escape angles.

All this didn't mean that the GPA or Seep (Sea Jeep) was a totally useless, uninteresting vehicle, but rather a machine with a limited use, adapted to specific conditions that weren't very common.

When these conditions were met, the GPA seemed to have satisfied its users.

Amongst its unique characteristics, the Seep had, as any boat did, a mechanical capstan. It was also equipped with two hold pumps. Finally, the screw shaft combined with the transmission case exit, gave it a rotation and therefore a speed proportional to the chosen gear and a reverse function in water. Also the wheels' rotation in water could be fixed according to the need for speed and propulsion.

ARMORED JEEPS

June 1942: The Willys/Smart T25 Jeep during the first tests in Aberdeen.

It was at the headquarters of SMART in Detroit that the first studies of armored Jeeps were given to the U.S. Army in 1941. SMART specialized in armored trucks for transporting money and also in armored cars for personalities all over the world.

Armor plating consisted of an armored radiator grille, hood and windshield in front and two armored half-doors. Not only did the extra weight noticeably reduce the vehicle's general performances, it also made road holding unpredictable. As a matter of fact, the enhanced center of gravity, partially because of the windshield, together with the additional strain put on the original leaf springs, caused worrying swinging movements on slopes and round bends. Unfortunately, this impression was confirmed by an accident

Fort Knox: The Army Force Board tryout the T25 E2; the chassis used is now the Willys MB.

Last evolution before the project was abandoned in 1943: the T25 E3 has total armoring and has diverse technological advances concerning inclined armor.

that claimed the life of a test driver during trials.

So, from 1942 to 1943, SMART undertook the successive study of four other prototypes, since then officially called T25. Three other prototypes were produced, the T25E1, the T25E2 and the T25E3. But finally, none of them was used because the weight, the engineers' unbeatable enemy, not only hindered the vehicle's performance but also put the mechanical parts under great strain, which posed reliability problems during trials. With the aim of reducing the weight and the enhanced center of gravity, the only solution was to change the armor plating which had to be thinner (from 8 to 6 mm) and not so high.

But when a good idea surfaces, it isn't dismissed so easily. It was therefore the troops in the field who resuscitated the armored Jeep. After the rather timid tentative carried out in December 1942 in Tunisia by the American troops, armored Jeeps appeared in Normandy in 1944 and particularly in the Ardennes at the end of 1944. From what we know, the modifications carried on in the field were mostly done by the American parachutist unit, and notably by the 82nd Airborne. The vehicle, imagined and

Last evolution of the T25 based on the Jeep MA's chassis.

December 1944/January 1945. Soldiers, probably from the 101st Airborne Division, using an official armored car in Bastogne, Belgium.

On the road, in spite of all the good will of its brave 2.2L Go Devil, the armored Jeep struggles under the weight of its shell; acceleration is slow and the back end slews from side to side around bends.

January 1945: At Apach, a small village in the Lorraine, an armored Jeep accompanies the infantry that in a few kilometers will cross the German border.

This armored Jeep used in the Battle of the Ardennes, was equipped with a twin bazooka, instead of a machine gun, mounted on an M31 support.

Redoubtable as a scout-car, the Jeep increased the vanguard troops' protection at the frontline.

modified by regimental workshops had a surplus weight of around 300kg and the armor plating protected the interior on three sides as well as the radiator grille. Obviously the armor had little in common with the famous treble equation that the SMART engineers, who tried perhaps too hard, had failed to overcome. The relatively basic armor covered at best the vital parts of the vehicle. As well, its low height reduced the center of gravity enough for the road holding to be acceptable and finally the relatively limited extra weight made the mechanical viability encouraging.

Once settled behind the armoring, the armored Jeep didn't give its occupants very much in way of a view, but this was the price of protection. Note the presence of a SCR300 radio set and an M31 gun carriage and the famous .30 caliber Browning.

Honolulu, 1945. This Jeep is being used to transport the wounded. Its markings indicate that it belongs to the medical section of 100th Infantry Battalion of the 34th Division.

MEDICAL JEEPS

This medical Jeep of the U.S. Marine Corps (built by the Australian company, Holden) is used to transport wounded to a Douglas C-47.

The existing medical vehicles such as the Dodge WC54 meant that the medical Jeeps we know from operations in Europe, only played a minor role, either as additional vehicles when there was a sudden influx of the wounded or when it was necessary to drive out to the Front Line or other places were access was particularly difficult. In such circumstances, it was naturally the Jeep, the real general help of the Army which performed this new service because besides its polyvalence, its speed and its cross-country qualities, it was also the most widespread vehicle within the fighting units.

No changes to the original model were made, so it was therefore once again because of improvisations that a number of modifications came about. With everyone acting according to specific needs and resources to hand, it can be seen that in spite of obvious similarities, there were almost the same number of different modifications as there were medical Jeeps. So according to the units and also the different armies: British, Canadian or American, several adjustments came to light. The most simple, but also the most precarious, amounted to the positioning (transversal or not) of one or more

This beautifully restored medical Jeep, proves that it was possibly to carry two stretchers thanks to some basic modifications. This was the configuration in which it was mostly used.

August 1942 in the U.S. Side view of a Jeep equipped with lateral stretcher supports, belonging to the 62nd Artillery Regiment.

A medical Jeep being tested by the 62nd Artillery Regiment during exercises in August 1942 in the U.S.

Overhead view of a Jeep fitted out to carry four stretchers. It belonged to a medical battalion stationed somewhere in England in February 1943.

During the Battle of the Ardennes, a Jeep belonging to a medical unit evacuates the wounded on a support fixed to the hood.

A medical Jeep circulating in the waters of the Moselle, transports the German wounded away from the battle and crosses a Jeep heading for the frontline.

stretchers on the hood and as many in the back. In such conditions, one can imagine the wounded believing that their troubles had only just begun! The seriously wounded had tubular structures that allowed the back overhang to be extended to take on two stretchers placed facing the front. Another tubular arrangement on two levels meant that three stretchers (two one over the other) could be transported, which meant that any eventual nurses had to cling onto the sides of the Jeep as best they could.

In every case, it was extremely uncomfortable, but in the circumstances what other choice was there?

Having being fitted out this way and with their basic canvas cover, which wasn't really suitable for evacuation of the wounded, all these homemade changes served no other purpose than to allow the Jeep to answer a given and momentary need, without inferring modifications appropriate for an exclusive and definitive medical use.

The only Jeep ambulance worthy of the name and constructed for that purpose was a version intended for the Marines in southwest Asia; about 200 examples were built according to the plans drawn up by the Holden Company (a subsidiary of General Motors) in Australia.

RAILROAD JEEPS

This rail Jeep, photographed outside the station in Messac, Brittany in September 1944, was aptly nicknamed "SHORT CUT."

Like a number of modifications carried out on the Jeep, the idea of adapting it for use as a railroad car, came from those who needed it, that is to say the users. Therefore most of the changes were system D and homemade as one can see from the large diversity of equipment to be found.

As far as we know, two types of railroad Jeeps were used during the Second World War.

The first consisted of the use of the Jeep as a replacement for a locomotive; the second responded to the tactical and logistical needs during close combat where the roads were unsafe and often overcrowded. In emergencies the Jeep could momentarily replace destroyed locomotives and act as a means of transport, thus ensuring supplies by rail.

Finally, as a purely tactical use, with the shifts and changes during combat, the Jeep could be put on the rails to

Like a real locomotive, a Jeep GPW on railroad tracks in Burma. The astute system mounted on the Ford's front fender, dribbles out fine sand to stop the metal wheels from skating, while it hauls heavy convoys weighing up to eighty tons!

serve as a scout car, for example, as the enemy's attention was usually turned towards the roads.

So in the first type of use, once fitted out with a bumper, a counterweight and a towing hook, the Jeep could tow a load of up to 100 tons! The only major problem when used this way was braking!

For the second type of use, the vehicle's general configuration was completely different.

In the case in point, it mustn't be forgotten that above all, the Jeep was a combat vehicle that had an originality and a versatility all of its own. Thus, when the Jeep was installed on the rails it always kept its original wheels sprouting out of the bodywork, as its versatility was its major advantage and the reason for its existence. To this end, the Jeep was often used on the railroad with its own combat wheels, either half or entire (although they weren't made for this use). In this case, it goes without saying that the small diameter of the wheel alone, greatly affected the vehicle's speed.

A rail Jeep pulling a wagon along a portion of track on Luçon Island, January 21, 1945. The charge towed by this Jeep is equivalent to six 2 ½ ton cargo trucks.

August 28, 1944. A rail Jeep towing wagons between Myitkynia and Moguang in North Burma.

6X6 JEEPS

This Jeep MT-TUG here with its canvas hood, built for the Marine Corps, is recognizable by its high sides.

It was in 1941 that the first Willys 6x6 Jeep was built. The MB ¼ ton 4x4 was only just in production when the Toledo factory had already started on a ¾ ton 6x6 project.

The first model consisted of creating a 37mm SP gun support, the *37mm Gun Motor Carriage T14* based on a *Willys Slat-Grille* (early MB) which met the needs of the *Tank Destroyer Command*, who wanted a fast self-propelled machine. The second prototype was also based on an MB, but the project didn't go any further.

Not wanting to waste the efforts put into the conception of this small 6x6, Willy-Overland carried on and created the *Willys MT-TUG, ¾ Ton, 6x6, Super Jeep.* The most probable explication behind its name was, *Military Truck* (or *Tractor*) and *TUG* for its towing capacities. This towing role was very exceptional; the same vehicle could serve for transport or as a semi-trailer tractor. For this last use a coupler, more often called a harness was installed, in the center of the body just over the double axle. The MT-TUG prototype came out at the beginning of 1942. Its Registration Number was 828993, the first "8" indicating that it was a light tractor. The Quartermaster Corps took the project under its wing and an order for 15 units was placed (Contract W-303-ORD-4623), listed as "Truck, ¾ Ton, 6x6 Tractor." Delivered in 1943 they were given Registration Numbers from 2179612 to 2179626, the "2" indicating a light truck. The MT-

The second version of the *T-14 37mm Gun Motor Carriage* equipped with a welded on gun shield, was photographed at Aberdeen, Maryland in 1942.

The Willys Super-Jeep, ¾ ton 6x6, MT-TUG harnessed to its cargo trailer especially built for airfields.

TUG had been tested as a light cargo truck and as a semi-trailer.

In this role it had been linked with a variety of trailers built by Fruehauf (cargo, workshop and ambulance).

During this time the U.S. Marine Corps, also interested by the idea of a vehicle larger than a Jeep but with the same mechanical elements, tested in 1943 another version of the Super-Jeep. The back part of the bodywork had higher sides and could also be used for various functions without modification; all the elements for diverse missions had been incorporated in the basic model: cargo transport and passengers (seven), mortar squads (four men and material), or ambulance (three stretchers). Created in three examples this U.S. Marine Corps model wasn't developed any further than the Army version.

To completely cover the subject, we should mention an experimental prototype: a sort of armored scout-car.

The T24, tested in 1942 at Fort Knox and in Aberdeen, wasn't developed any further either.

Even though, generally speaking, the tests had been satisfactory, the project was an ambitious one. With the *Willys MT-TUG, ¾ Ton, 6x6, Super Jeep*, the company in Toledo was trying to push its way into a number of spaces already occupied by the big names of the U.S. automobile world: Dodge (transport and ambulances), International Harvester (Marine Corps version) and White (Scout car). It was a strong coalition that was impossible to stand up to.

The MT-TUG was also tested in 1942 in its medical version for transporting three stretchers and intended for the Marine Corps.

The final version of the Willys-Overland "Super Jeep 6x6, 37mm Gun Mount" had a cover whose framework was arranged on the back of the sides.

ELONGATED JEEPS

This Jeep, extended by 76cm came out of the regimental workshops, as no serial modifications were adopted. Notice the addition of an extra cover arch as well as two rows of seats.

Only one project baptized ML-W4 was undertaken by Willys in 1944 to create a heavy ½ ton Jeep. The aim was to bring the useful loading weight up to 500kg. The wheelbase was enlarged to 2.33m (30cm more) to take a 7.50x20 tire identical to the GMC. As this sector was already covered by Dodge, it wasn't followed up.

The superb restoration of this elongated Jeep "Purple Heart" shows the diversity of the transformations that the Jeep underwent. Apart from a few examples, unfortunately, these transformations went no further, because the role as already filled by the Dodge WC51 and WC52.

Elongating the body meant that eight soldiers could ride in the Jeep. Its rigidity forced the Jeep to keep to beaten tracks.

The American Coast Guard used elongated Jeeps so that they could carry ten men. These Jeeps had larger wheels so that they could move easily through the sand.

Another modification was done for the U.S. Coast Guard, who patrolled the shores, so that they could transport a squad of ten men. This Jeep, whose chassis was lengthened by 30cm, still had two axles and four-wheel drive. Eight men could sit in the back on benches fixed along the sides.

It had large balloon wheels, making driving across the sand easier.

Other homemade modifications were carried out directly in the field. This was the case with the U.S. Army Air Force where a few Jeeps were elongated to carry entire B-17 or B-24 aircrews.

This was the only modification done in the field to be officially recognized. These Jeeps could carry ten men sitting crosswise.

Another well-known homemade modification was the one made by the men of the 39th Signal Construction Battalion of the 9th U.S. Army from two crashed Jeeps.

This elongated Jeep was built from several vehicles. It could transport ten men. The "X" following the serial number indicated that the vehicle was built from several others.

SAS JEEPS

This redoubtable SAS Jeep (Special Air Force), immediately gives the impression of incredible fire-power. Equipped with two twin-barrel Type K Vickers in the back, and a single barrel in the front, a Browning .50 caliber could also be added.

Who Dares Wins!

It was in 1941 that David Stirling, a Scottish officer in the British Army, had the idea of forming Strike Squads consisting of four or five highly trained and specialized men, intended to infiltrate behind enemy lines to destroy predetermined objectives. Equipped with Jeeps specially prepared for the numerous raids and above all formidably armed, the SAS (Special Air Service) created havoc among the enemy ranks! Stirling was sure that instead of excessive numbers, a small unit of highly specialized men would undoubtedly have a better chance of success as they would be less noticeable. They would work directly behind enemy lines.

To facilitate cooling in high temperatures, these Jeeps had their radiator grilles removed and an expansion tank was added.

Even if putting the idea into action started off on the wrong foot, due to the resounding failure of the first SAS squad in November 1941, mainly caused by atrocious weather conditions, the series of raids that followed had a certain success. They had learnt their lesson from the first failure and the men spent each day getting used to their desert environment and working out operation methods.

It wasn't until the arrival of fifteen Jeeps that a real revolution took place, philosophically and in method. It didn't take long for the SAS to realize that the combination of the formidable resources of this new vehicle and the strength of adapted arms, made a more effective weapon than a dagger. Also, leaving the combat zone in a Jeep was more comfortable and more efficient than it was on foot.

In the North African desert, SAS Jeeps were severely tested. The intensive use and excess weight quickly gave a vehicle a very misleading worn out look.

January 18, 1943. A patrol comprising three SAS Jeeps preparing to leave on a mission, under the command of Lt. McDonald. SAS founder David Stirling is at right.

So that no one died of thirst – the vehicle or the crew – the SAS Jeep had a large stock of jerry cans.

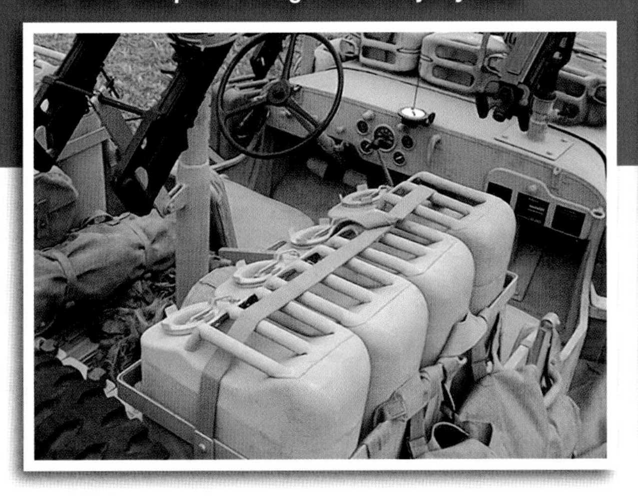

Straightaway the SAS lost no time in modifying their new means of transport, starting with the essentials: arms. Plane carcasses scattering North African landing strips are still a reminder of the "raids" carried out by Stirling's men who came to relieve them of their Vickers Type K or caliber .50 machine guns which made up their weaponry.

With another caliber between 7.7mm and 12.7mm associated with high firing rates, the firepower was clearly there. The lightweight of these weapons was also appreciated, especially by the Jeep. Finally, concerning firing durability, there wasn't better on the market; because they had initially been built for planes, these arms didn't have cooling problems.

The SAS Jeeps had a large number of modifications. Among the most noticeable, after the weaponry, was the increased number of jerry cans – both for water and for gas. A standard equipped Jeep could carry up to fourteen: five on each front wing and four extra ones laid flat and fixed to the hood, giving a total of nearly 280 liters and weighing accordingly!

In front was an over-large expansion tank to capture any loss of cooling liquid.

The radiator grille bars, except for the two central ones (for rigidity) were sawn off to make cooling the motor via the radiator easier.

Thanks to this new tool, the British Special Forces could finally dare to attack objectives far behind enemy lines, accentuating the element of surprise. It was precisely because the German commanders didn't expect this type of action that made the raids such a resounding success. It didn't take long for the SAS motto to become "Who dares wins!"

FRENCH SAS JEEPS

Soldiers from the 2nd *Régiment de Chasseurs Parachutistes* who retook Saint Hubert village in the Ardennes.

The French Forces greatly participated in the constitution of Stirling's training staff, and were therefore among the first to appreciate these units' capacities and in 1944 they took the step of creating two SAS regiments of their own: the 2nd and 3rd *Régiment de Chasseurs Parachutistes.* Both of these regiments were organized around four combat squadrons each counting, other than a command group, four fighting platoons, each of which had three Jeeps; a regiment totaled nearly sixty armed Jeeps.

They therefore took part in numerous actions during the campaign in France and long afterward.

Although most of the SAS units were disbanded after the war, their organization methods and their acts became the basis on which Special Forces all over the world were built, even to this day.

1944: For the regiments of the *Régiment de Chasseurs Parachutistes*, this SAS Jeep was equipped with three 7.7mm Vickers K and a Bren machinegun. In this particular case, they had considerable firepower.

A port in Great Britain. These Jeeps are aboard an LCT preparing for the Invasion. They are equipped with Deep Fording Kits and M51 gas-detection paint in the star's circle. Their markings indicate that they belong to the Commandment Company (HQ-19) attached to the General Staff of the 1st Infantry Division (1-X).

CALL OF THE WHEEL

Numerous Jeeps were stocked in the California desert before being delivered to the 3rd Armored Division on October 22, 1942.

Let's not draw out the agony any longer and succumb to the wonderful 1944 Willys MB's invitation. Once perched over the gas reservoir and settled behind the imposing three-branch steering wheel, all we need to do is to start the engine.

It's not until the key is turned and after a few revs that the right foot activates the foot starter. Straightaway the motor starts purring happily. The sound of the good old four on-line cylinders is immediately characteristic. The basic exhaust line filters make a sound both hoarse and peaceful as only the Americans know how to make it. Even if the song is still more melodious on the exterior, it constitutes in fact, once you are behind the wheel, a likable companion that gives more information than the rev-counter.

Sixty years after its appearance, the Willys Jeep remains unique. A simple glance at it gives you a desire for freedom. Sitting behind the wheel is a magical moment.

But it isn't until the Military 600x16 swings into action that the real sensations begin with a stimulating effect that goes straight to the head.

Only having three gears when it could have accommodated more, the third is logically high range, giving a smooth ride. The other two are low range meaning that they give it vigorous boosts seeing its good old 2.2l power. The reverse, the lowest gear has a certain nervousness on hard, dry ground and picks up power on soft ground.

The real advantage comes from the low speed availability that the generous capacity of the Go Devil knows well. On the road as well as on rough terrain, this offers a flexibility of use that economizes perpetual gear changing. The first gear as well as the reverse gear, unsynchronized, should only be engaged while the vehicle isn't moving, to avoid making the gearbox scream. The others don't present any problems even when changing down. Whatever one may say, a Jeep gearbox, if it is in good condition, doesn't need double clutching techniques.

After a few miles, we can assess the vehicle's road performance, and it must be admitted that the Jeep is a long way from our current models. But when we take into account is original purpose as a fording vehicle and its ancient age, we must admit that it adapts to the asphalt brilliantly.

Another thing that we notice is the tendency the Jeep has to over steer, caused by the structure of its propulsion and its relatively light back end. This problem is accentuated on a wet road where the Jeep's back end swings to one side on successive bends taken at steady speed. We should also add that the situation isn't helped by the "Military" tires which have narrow threads.

The dashboard (if you can call it that) is particularly well equipped for the epoch. All the information about the motor is grouped around the speedometer (in miles). The temperature gauge, the oil-pressure and the fuel gauge are part of the standard instrumentation in all Jeeps.

To brake the ton of khaki, the Willys Jeep has four hydraulic drum brakes. If these are well adjusted, in good condition and dry, the Jeep, in spite of the legend, does actually brake – depending on certain conditions!

There are actually two. The first one comes from the non-assistance of the braking circuit which therefore means putting frank pressure on the pedal. The second comes from the simple fact that the braking circuit is decentralized to the left side of the vehicle, which means that when braking the brakes on the left side come into use before those on the right side, resulting in the Jeep's swinging

September 11, 1944. In the sector around Luca, Italy, American soldiers evacuate nuns from the Front Line, in their "slat grille" Jeep.

slightly to the left when the brakes are applied. You need to anticipate this and adjust the wheel accordingly towards the right.

Obviously these precautions don't do much for driving comfort, but for a driver who knows the problem and is used to it, braking is efficient as long as you avoid doing it when fording a river or driving through a quagmire.

So, if in spite of a few admitted handicaps, the Jeep handles well on the road, you can guess that this isn't its preferred terrain. Its principal qualities only really stand out as soon as its "military" wheels hit dirt. Anything else is just a starter.

On its home terrain the legend says that. "the Jeep climbs trees." If this metaphor is somewhat exaggerated, it isn't too far from the truth. The Jeep quickly shows itself to be a cross-country vehicle par excellence. It has climbing talents that can be seen on hills with a 60% gradient according to the type of ground.

It is helped in this by equipment which, put end to end, makes a real forward motion machine. Here the wheels, so clumsy on the road, keep all the promises of their shape. Besides, once the front axle and the reducer are engaged, a tremendous "tractor" effect is immediately felt: the top speed of each gear is cut in half but the power reserve produced seems inexhaustible. Nothing can stop the four wheels from turning. It remains to be seen whether their rotation goes in time with the vehicle's movement. Only the nature of the ground is the master of the subject – or almost. In fact, at this stage, only certain parameters still could tip the balance. Among these we note the ground clearance and the weight, two areas where the Jeep doesn't need any lessons.

When it's stuck ... it's stuck! Sergeant James Rogers tries to get out of a quagmire in the Nancy area in November 1944. He can only hope to get out of this mess thanks to the Jeep's power and lightweight.

On the downhill side, if the brakes aren't always the best method of stopping (especially if they are wet), the engine braking, especially at a slow speed, is redoubtably efficient. It enables the four wheels to harness and control the speed downhill, without even touching the brake pedal. All you have to do is to manage the descent by a simple pressure more or less important on the clutch pedal.

If its performances in cross-country are plain to see, the Jeep also knows how to pick up the pace without losing any of its sufficiency. Without the technical refinements of modern suspension, the old leaf springs offer a reasonable flexibility and ensure a uniform contact with the ground. This small handicap is accepted by the driver whose discomfort isn't eased in any way by the softness of the seat's cushion. But after all,

In October 1944: an official Signal Corps photo showing the Jeep in a tough situation. With a helping hand, it will get out of this mess thanks to its excellent 2.2L capacity "Go Devil."

wheel at least offers the advantage of being able to stop at any time to ease the torture if it hadn't been transformed into a fun rodeo.

On hilly tracks, as on the road, even while driving in normal mode (propulsion), the Jeep has a tendency to over steer when going round bends, which can be pretty funny if this is the sensation you are looking for. However, for more security, engaging the front axle corrects this and prevents the back of the vehicle from sliding to the left when going round bends. The wheel's narrowness and the vehicle's "light weight" which add to the problem, will dictate the vehicle's speed in a bend.

The only real snag, which is progressively felt when in third gear with the front axle engaged: the 2.2L quickly reaches its limits on soft ground and struggles to pull along its four wheels simultaneously. It is therefore necessary to engage a lower gear that penalizes in turn the speed.

One is forced to admit that a strange blood runs through the Jeep's tubes giving it a rich character full of subtle alchemy which make it a living vehicle, energetic and bubbly, always amusing and fascinating to drive and whose cross-country capacities seem endless.

But if it was necessary to remember just one thing, know that the Jeep is undoubtedly the vehicle the most capable of taking you from one point to another no matter what road you choose, whether it's over a mountain, through a valley ... or across German lines.

Even if it's not its cup of tea, the Jeep adapts to the road without any difficulties. This versatility is due to its agreeable motor and its lightweight that makes it lively and sharp.

A Jeep crosses the border between Belgium and Germany on October 7, 1944.

BUYERS GUIDE: HOW TO ACQUIRE A PIECE OF THE LEGEND

If thousands of examples of the Jeep were built, a lot of them finish their lives like this one, which belonged the *1ère DB Française* and which was blown up by a landmine.

Hallelujah! You have decided to offer yourself a part of the myth and buy a Jeep. It certainly wouldn't be us who will try to dissuade you. Know moreover, that your future Jeep has every chance to make your automobile capital prosper as well as, if not better than a savings account or any other form of open-ended mutual fund. It will also give you pleasure and sensations that you'd never discover in investment schemes! It's something so rare in the automobile world that it should be mentioned.

Thanks to specialized literature, the Serial Number is the best way to identify a Jeep's origin (here, a Hotchkiss M201) and to give it an exact date.

This is because, unless you get a pig in a poke or royally cheated, it is customary to sell your Jeep for a higher price than you paid for it (as long as you have kept it in good condition, which isn't always the easiest thing on earth!). This well-known phenomenon is explained by a simple macroeconomic rule, made clear by J.M. Keynes in his time – that of offer and demand. In fact this last never stops increasing due to a feeling of affection, to the Jeep's classic appearance and to its history that every self-respecting journalist never lets anyone forget. Up to this point, we can only be happy about this pleasing fact.

However, it's on the offer side that the problem lies, because you have to face the fact the finding a Jeep is becoming increasingly rare. Over sixty years of good and loyal service (and hard labor for some of them) is very wearing. And to aggravate matters, the enormous supplier that represented the French Army, through the famous sales of domains, has recently closed down because the reformed stocks are now exhausted. The only way to unearth a real treasure is through private individuals or professionals who specialize in this vehicle.

To point you towards the model you want and to be able to discern the differences between American and French Jeeps, if that aspect worries you, we urge you to consult specialized literature (see the bibliography) because the differences are numerous and often insignificant for a novice. You should also know that the prolonged stay with the veterans of the French Army of World War II has only complicated things. Almost all of them were in fact (for their own good) rebuilt in ERGM *(Etablissement de Réserve Générale du Matériel)* workshops, where they had pieces and parts attached that came from a stock of odd bits and bobs. Therefore it isn't uncommon to come across a Jeep that has a Hotchkiss body welded to a Willys or Ford chassis ... or the other way round.

Historically, this state of affairs might be more coherent and status enhancing than trying to reconstruct a Jeep that is 100% American as none of the various made or copied pieces would have a common past. Each Jeep has its own history (both during and after the war) which in itself is a real bonus and it would be a shame to want to deny it. But obviously, this is just a question of taste.

Finally, for your future research, remember that any self respecting amateur or DIY-er usually keeps a file of photos showing each stage of the restoration which will give you an idea of the quality of work carried out.

The Price of Fame

Price wise, the Jeep whatever its model (Willys, Ford or Hotchkiss) is a victim of inflation like any other product of value and hard to come by. But as to the price you'll have to fork out to buy the Jeep of your dreams, this will depend on the mark and the quality of the restoration.

So a Hotchkiss M201 in good working condition and presentable would be difficult to negotiate for under 9,000€ and could easily reach 14,000€ if it is perfectly restored. On the other hand, for a Willys MB you would easily have to lay out the trifling sum of 11,000€ to have any hopes of finding yourself behind the wheel of a vehicle in good working condition, as long as you don't look too closely at the origin of some of the parts. The price increases as the quality of restoration and the origin of the parts is specified. So don't be surprised to find a superb specimen being negotiated at around 15,000€ and some reach 20,000€ when their rarity value (first models) or their museum status is justified. Finally, if Henry Ford is your idol and you wouldn't dream of driving any other vehicle than one of his bearing the seal of the Detroit giant, you should know that in an identical condition, a Ford GPW is negotiated at between 1,000€ and 2,000€ more than a Willys MB. The reason for this is because of the less widespread distribution of the Ford model and also because of Ford's identification policy of having the famous "F" stamped on the different parts, so dear to a collector's heart.

Offers are split into two groups: individual sellers and professional sellers. Obviously the latter's prices are somewhat higher than an individual seller, usually because of the quality of restoration carried out to be able to offer a guarantee. Buying from an

On the Willys MB Jeeps, the serial number is stamped on a zinc plate and riveted to the front left joist.

On the Ford GPW Jeeps, the serial number is stamped directly onto the front left joist inside the motor compartment.

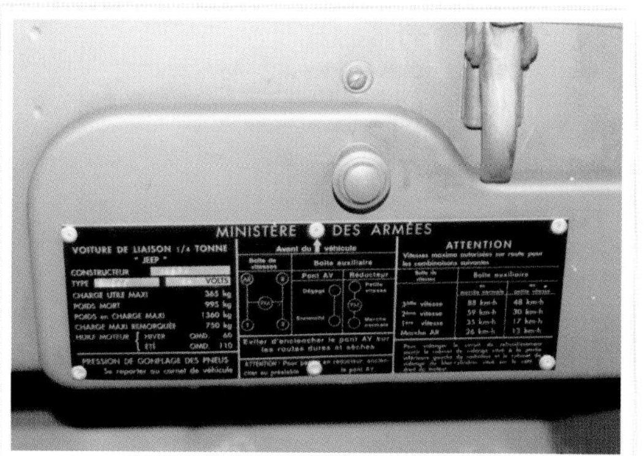

Each Jeep has constructor and instruction plates fixed on the glove box door. The Willys and Ford Jeeps have three screwed on plates (R4). Only the Hotchkiss plates are affixed with rivets (R5).

individual seller can be a good option when you know for sure exactly what restoration has been carried out and its quality or more simply because you aren't put off by DIY!

The Administrative Circuit

The vehicle inspection is an obligatory passage before you can register your future Jeep, whether the vehicle documentation is normal or collection. If DIY isn't your cup of tea, don't hesitate to go to a specialist for repairs or risk coming away from the prefecture empty-handed and having to wait before being able to play with your new toy.

As to administrative costs, they are subjected to thirteen horsepower. However the bill is divided by two as with any other vehicle of over ten years old.

Reviewing the Machine:

Who is Who?

A part from your erudition and omnipotence about everything concerning the Jeep, it's the vehicle documentation that will give you all the details about the pedigree of the Jeep you long for. This is the only official document that attests to the exact identity of an automobile. It should specify the mark: Willys, Ford or Hotchkiss, as well as the Jeep's chassis number. If, however, you want to protect yourself from unpleasant surprises and squash any possible doubts about the origin and nationality of your Jeep, we highly recommend that you refer to specialized literature (see bibliography), the only way to identify with certitude the origins of the chassis, the bodywork and the motor. To be on the safe side, don't put too much trust in the date the vehicle was first registered noted on the vehicle's documents to put an accurate age on your Jeep, except perhaps for a Hotchkiss. You would do better to put a date to it by the chassis number (once you know what it is), once again referring to specialized literature. One last piece of advice: Europe hasn't yet decided on the rights to drive "collection" vehicles in the future; therefore it would be better to have a normal vehicle registration document rather than a collector vehicle registration one.

Throughout the campaign in Europe material often got broken and the administration couldn't keep up with sending out spare parts. To repair their vehicles, everyone fell back on the good old "system D." This soldier is looking for a wheel in a depot full of damaged vehicles.

Bodywork

A visual jewel case, the bodywork survives time's outrages in various ways depending on the model, its age, and the way it was used and stocked. No matter the whether the body is American or French, it will obviously bear traces of its past tumultuous life and will therefore rarely be lucky to emerge without a mark on it. Repairs are completely normal and convenient to help it keep its brilliance of long ago. It is up to you to verify the quality of these repairs.

While we are on the subject, the footwells and the tool storage space are the real Achilles' heel because they are the first places that damp attacks if the vehicle hasn't been stored properly, so don't forget to check these areas carefully and if necessary, check out the quality of the repairs. The reservoir's protection space isn't much better off and if there is any perforation it will be on the level with the opening of the axe's housing or at its base.

You should also carefully inspect the famous wooden reinforcement crossbar that runs under the body, encased in a metallic U. Because of its composition it has a tendency to rot (almost systematically with Ford and Willys) and retain moisture that causes it to swell the floor or to pierce it. If this is the case you'll have to judge the appropriate repairs carried out.

Finally, on the identification side, it's unfortunately becoming increasingly difficult to be sure the bodywork is Ford's or Willys' because faithful copies made in Asia are available on the market. They even go as far as recopying early Willys or Ford's first series of models, which bear in either case the mark stamped on the back left hand side. Besides the subtle dissimilarities (resulting from the different production apparatus) obvious to the most experienced collectors, the simplest way to unmask a copy is to examine the condition of the bodywork. An original Ford or Willys is unlikely to have a body as new looking as a freshly painted Asian body, except in case of filling, which is not necessarily a bad thing when it is practiced with economy and correctly. Another give away is that on a copied model the sheet metal is usually thinner than on the original body.

This plate welded to the inside of the chassis' right joist on a level with the back axle, indicates that the Jeep was repaired by the ERGM works at Maltournée. The repair date is supposed to be stamped on a plate and riveted to the dashboard.

Each constructor (Willys, Ford, Hotchkiss) turned out their own small dissimilarities according to their specific production apparatus. The bodywork reinforcement is an example.

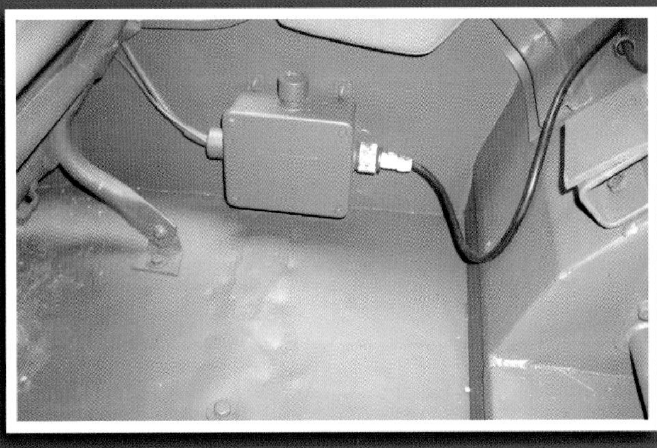

Each constructor (Willys, Ford, Hotchkiss) turned out their own small dissimilarities according to their specific production apparatus. The bodywork reinforcement is an example.

Engine

The cylinder block and headset in good old cast iron doesn't help the "Go Devil" keep cool in hot weather. But apart from this small weakness that betrays its age, we must admit that this venerable piece of mechanism is unbreakable and will take you to the end of the world, as long as you repay it with a minimum of maintenance. For the more skeptical among you, know that no less than six borings out were foreseen to avoid the cylinder bores from roughing up with wear. Try to find out how many times the engine block has been bored out so that you will know how many times you can have this operation redone if necessary.

You should also know how to interpret the exhaust pipe's language because persistent white smoke usually means that the head gasket's sealing

The tool storage space is another place where rust sets in and causes a lot of damage.

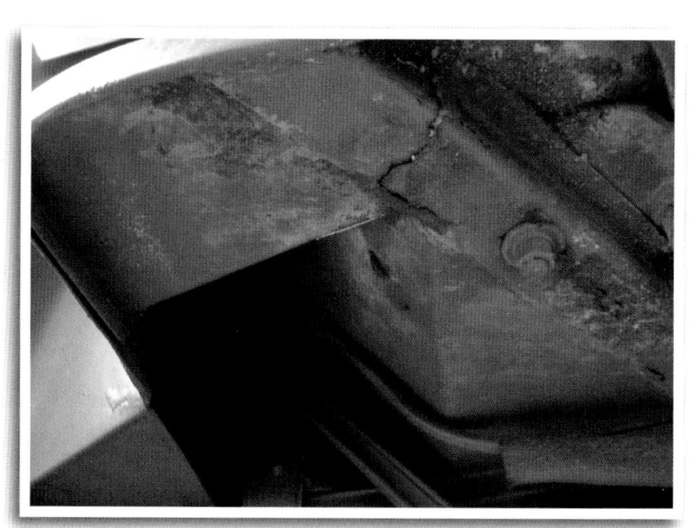

The famous reinforcement wooden crossbar encased in a metallic U is undoubtedly the Jeep's best known Achilles Heel, which with help from damp pierces the bodywork and the U.

is wearing out and when the smoke turns bluish it's time to think about changing it; a motor in good working order doesn't emit smoke, except when starting up.

Before buying, listen to the "Go Devil" to make sure it's running smoothly, especially when it's hot. Carefully check for leaks under the motor sump and under the transmission and transfer boxes, although slight oozing in these areas is normal.

Transmission

The engine's muscles, which transform on the ground the energy supplied by its generous 2.2L heart, they have the resistance of a long distance runner. Like the drive shafts, they have an unwavering endurance as long as you don't ask for a sprint! The drive shafts are fragile because they are undersized. So we don't recommend trying burnouts on dry and tarred roads or you will risk seeing a back wheel drive shaft twisting like a dishcloth that's being wrung out. If this happens you could hope to limp to a repair shop by engaging the front axle.

On the transmission box side, remember that the first gear isn't synchronized and grinds unless the Jeep is stationary. The second has a tendency to pop out of gear in certain models and often needs having the distorted fork replaced. The third gear also grinds in some cases. This isn't a serious problem in itself and there are two solutions to the problem: either you have the transmission box adjusted, or you get used to changing gears gently.

One of the Ordnance Corps' numerous depots. All damaged or broken-down vehicles were stocked in enormous depots so that they could be cannibalized. All the working parts were dismantled and the rest was simply dumped together, as this photo shows.

Brakes

It's better to say straightaway that this is not part of the Jeep's fatal charm! However, the drum brakes are more or less efficient at stopping the ton of khaki on the road, but are quickly inefficacious when fording or driving through a really muddy terrain. Also, it is normal that the Jeep swings slightly to the left when braking; it's a characteristic symptom of the off centering the hydraulic circuit to the left. Sorry to have to tell you, but you'll have to get used to the trick of righting the wheel to offset this problem.

Electricity

If mechanical comfort and tranquility are your first preoccupations, forget original Jeeps equipped with 6-volts, as they are synonymous of capricious starting up. In fact the six little volts carry wheezy power to the starter which is insufficient if the ignition setting isn't perfect or if the gas pump doesn't do its job straightaway. However, if all is properly adjusted, a 6-volt Jeep will start up correctly under any conditions, but it doesn't like approximate adjustments or shoddy maintenance. To get the Go Devil up and going, you should remember to prime the gas pump first each time the motor is cold. Because, unlike the M201 24-volts, it doesn't have a circuit breaker, you must remember to unhook the battery if you're not going to use the vehicle for any length of time that exceeds a week, otherwise it will discharge. Another thing to remember – you have the original lighting, in other words, less powerful, which needs to be fed by electrical wiring in perfect condition.

Life seems so much simpler behind the wheel of a 12- or 24-volt Jeep, which are closer to today's standards. The power sent to the starter will forgive a lot of ignition adjustment errors and will relegate certain operations, such as priming the pump, to the rank of old customs. Most 24-volt vehicles have protected ignitions which means no more problems in damp conditions.

Steering

In this area as well, the Jeep shows its advanced age and power steering is sadly lacking. The result: the steering box, the joints and relays wear out and become loose. All this makes the steering a bit woolly but this stabilizes with time. So, unless you have to turn the steering wheel half way round before the front wheels decided to turn, you usually get used to this small handicap (as with other small details in the Jeep).

The motor, even if it isn't as clean as this one, is easily accessible, easy to tinker with and has a reputation for robustness if it has regular maintenance.

The transfer case along with the gearbox has the same reputation for robustness as the motor. The gearbox should be handled gently out of respect for its synchros.

The rear axle, like the front axle, doesn't present any major problems, if it is properly lubricated. However, be careful of the axle shafts that are rather fragile.

Suspension

If you notice that the left side of your Jeep is sagging a little, don't panic. This happens with normal utilization as the leaf springs on the left wear out faster than on the right, because of the extra weight of the gas reservoir and the driver. A normal Jeep has a slight tendency to tilt towards the left. If you feel that it is leaning way too far over, compare the leaf springs on each

The bar joints and direction levers are other areas that should be well greased and shouldn't be loose. This also applies to the steering box.

As for the suspension, it's a good idea to keep an eye on the leaf springs on the driver's side to see if they are wearing thin; at the same time make sure that the shock absorbers aren't leaking.

side of the vehicle and don't hesitate to replace the most deformed ones, or else you could have an unpleasant surprise concerning road holding, especially when going round bends.

Spare Parts

On this subject, know that your Jeep should grow old alongside you without too much difficulty, for the moment at least.

In fact, whether it comes from an old U.S. stock, old French Army stock or even a vehicle made in Asia, you will be able to find among this motley crew of suppliers, the spare parts that you need.

Be a bit careful about copies made in Asia whose quality is less reliable than American or French stocks, at least when it comes to replacing essential parts, especially driving and vital motor pieces.

And the price of all these marvelous parts, you might well ask. Don't worry about your wallet, as most of these items are relatively reasonable, as long as you can resist the hypnotic charm of original American parts, which are more expensive than gold these days.

Now that you have become a wise old man of the Jeep, all you have to do is look carefully through the newspapers, magazines and Internet sites to find the Jeep of your dreams. We wish you luck and hope that these few lines will help you in your noble quest.

The Hotchkiss 24-volt M201: the motor parts are placed differently. Because of the two 12-volt batteries, the air filter is on the front left wing and the regulator is up against the side of the body. Note the ignition as well as the wiring and the shielded spark plugs. The carburetor is made by Solex rather than Carter, the horn has lost its duct and the water-pump belt has been doubled.

BIBLIOGRAPHY

BIBLIOGRAPHY

Jeep, sur les traces de la legend, David Dalet et Christophe le Bitoux. ETAI 2003.
Jeep, Emily Becker and Guy Dentzer, 1994.
Only in a Jeep, Christophe Routier, Hermé, 2000.
La Jeep: un défi du temps, Jean Gabriel Jeudy, EPA, 1993.
M201, Pat Ware, Warehouse Publications, 2000 (English version).
Essential Military Jeep, Graham Scott, MBI, 1996 (English version).
La Jeep de mon père, Patrick André, ETAI, 1998.
Jeep Story News.

National Canadian Archives, NARA (National Archives and Records Administration). U.S. Army Signal Corps Archives. Military History Institute Archives (Carlisle). Patton Museum Archives (Fort Knox). Headquarters Museum Archives (Fort Eusti). Transportation Museum Archives (Fort Lee). Christopher le Bitoux, Jim Allen, CMIDOME, Ian Coxshall, Matthieu Dadillon, Yves Debay, Joseph Delecolle,, Cyril, Indiancars, De Platers, David Dalet, Jean- Pierre Françoise, Collection Patrick Sarrazin, Collection Didier Andres.

ACKNOWLEDGEMENTS

We would like to thank Indiancars for lending us their vehicles, for their photographs, and for all their historical and technical information.

GERMAN VISOR CAPS OF THE SECOND WORLD WAR

Guilhem Touratier and Laurent Charbonneau. This full color illustrated book covers the highly collectible World War II era German visor cap. Officer and enlisted caps of the Heer, Luftwaffe, Kriegsmarine and Waffen-SS are shown in over 360 high-quality photographs, and described in detailed text.

Size: 9" x 12" • 366 color photos • 80 pp.
ISBN: 978-0-7643-4458-9 • hard cover • $29.99

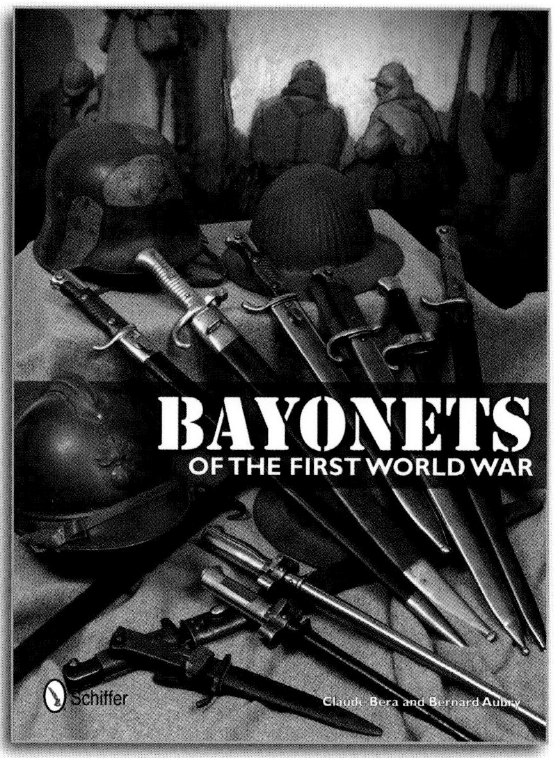

BAYONETS OF THE FIRST WORLD WAR

Claude Bera and Bernard Aubry. This detailed, concise look at WWI bayonets presents their development and use by all combatant countries. Images show full views of bayonets and accessories, and a clear look at hilts, connection points, blades, and a wide variety of manufacturer's markings.

Size: 9" x 12" • 273 color photos • 80 pp.
ISBN: 978-0-7643-4459-6 • hard cover • $29.99

Schiffer books may be ordered from your local bookstore, or they may be ordered directly from the publisher by writing to:

Schiffer Publishing, Ltd.
4880 Lower Valley Rd.
Atglen, PA 19310
(610) 593-1777; Fax (610) 593-2002
E-mail: Info@schifferbooks.com

Please visit our website catalog at *www.schifferbooks.com* or write for a free catalog.

Printed in China